Dams, Dynamos, and Development:

The Bureau of Reclamation's Power Program and Electrification of the West

By
Toni Rae Linenberger
with initial assistance from Leah S. Glaser

United States Department of the Interior
Bureau of Reclamation

For sale by the Superintendent of Documents, U.S. Government Printing Office
Internet: bookstore.gpo.gov Phone: toll free (866) 512-1800; DC area (202) 512-1800
Fax: (202) 512-2250 Mail: Stop SSOP, Washington, DC 20402-0001

Lifting the rotor on unit 4 in Shasta Powerplant with a 460 ton rotor lift. Central Valley Project: Shasta/Trinity Division, California. Date: Unknown. Photographer: Tom Papowich.

Interior of Canyon Ferry Powerplant. Pick-Sloan Missouri Basin Program: Canyon Ferry Unit. Date: May 7, 1963. Photographer: T.R. Broderick.

I am honored to be serving the Bureau of Reclamation as Commissioner during our Centennial. As a career Reclamation person, it is a thrill to serve as the nineteenth head of Reclamation and to be leading our organization into its second century.

I am particularly pleased that *Dams, Dynamos, and Development* provides the historical perspective on how Reclamation's hydropower program was started, evolved to serve millions of customers throughout the West, and is continuing to change to meet the challenges of a deregulated electric utility environment.

Reclamation's hydropower program is a centerpiece of our portfolio of services for citizens in the Western United States. Our fifty-eight powerplants provide electricity for over nine million Americans. The responsiveness of our hydropower generators provides a cornerstone in assuring system reliability in the West, in turn assuring that our customers will have a reliable supply of electricity. The power program also serves as a financial cornerstone, providing financial repayment assistance to many of our multipurpose water resource projects. Our water projects provide a reliable water supply for irrigation and municipal and industrial customers. Many of our power facilities also are home to tourism and recreational opportunities for visitors from throughout the world.

As the second largest hydropower utility in the United States and the largest provider of water in the West, millions of Americans depend on Reclamation power and water. I take this responsibility seriously, and I am looking forward to continuing our tradition of excellence into our second century. I am proud of our heritage, and we can look forward to a challenging, exciting future for the next 100 years.

John W. Keys, III
Commissioner
2002

Due to high flows coming into the Colorado River, the hollow-jet valves and the forty foot diameter tunnel spillways were used in combination with the powerplant to pass as much as 115,000 cubic feet per second of water. Colorado River Storage Project: Glen Canyon Unit, Arizona, Utah. Date: May 9, 1984. Photographer: Tom Fridmann.

It was December 1964, and the family station wagon had left the cold, snowy roads of Wyoming and reached the dry, warm desert near Glen Canyon. The sun was setting, and the new bridge gleamed copper in the sunset. We rounded the corner, and I caught my first glimpse of Glen Canyon Dam and Powerplant. To my fifteen year-old eyes, the sight was breathtaking. As we stood overlooking the dam and powerplant, I think I fell a little bit in love with them.

Little did I know, standing there in the brisk December evening, that some thirty-five years later, I would have the privilege of managing the Power Resources Office for the Bureau of Reclamation. I am still in love with the power program . . . the smell of cold, damp concrete in contrast to the warm, machine-scented air in the plant; the reassuring hum of turbines running and the rumble of water in the penstocks; the magical sight of a tidy row of generators, shafts turning, shining, generating electric energy to serve the West; the friendly faces and competent, professional work of Reclamation's power employees; the pride a plant's staff takes in telling me about their facility; the appreciation of our customers . . .. Each of these is indelibly etched in my mind.

I still get excited when I visit a Reclamation power facility. Each of our fifty-eight powerplants has a unique personality and contribution to the electric system in the Western United States. Taking a walk through our various plants is taking a walk through history. It is a delight to discover each plant – from the small, quaint, early-day plants to our sleek, modern behemoth plants. Reclamation's power program continues to serve the West as an exemplary model of public stewardship in the face of competing demands and a rapidly evolving regulatory environment. I hope you enjoy *Dams, Dynamos, and Development* and the colorful history of Reclamation's power program . . . and that sometime you will have the opportunity to visit one of our powerplants and experience its magic.

Deborah M. Linke
Manager, Power Resources Office
2002

Hoover Dam at night: then and now. The face of the dam is illuminated by a new metal halide lighting system, while the powerhouse wings are illuminated with the original high-pressure sodium lights. The first of the new lights was installed May 21-24, 2001. Boulder Canyon Project, Arizona, Nevada. Date: May 25, 2001. Photographer: Andrew Pernick.

Acknowledgments and thank yous are a very personal endeavor – not unlike the writing of a manuscript – even one undertaken in some sort of official capacity. A manuscript reflects larger ideas; at least one hopes this is the case. The kudos, however, recognize those who made the final product possible. Some of these people went out of their way to help me, and I greatly appreciate it. These, then, are mine: and very personal ones at that.

Michael Roluti, former manager of Reclamation's Power Resources Office and current Director of the Technical Service Center, agreed to this project in the first place and championed it throughout the writing and rewriting process. Mike's enthusiasm jump-started this project, and his vision kept it going and alive, as well as provided a definite shape to the final product.

Harvey Boyce patiently "taught school" to a curious historian and then provided a wonderful tour of Hoover Dam to show me how it all worked; as well as, patiently answered numerous, varied, and often esoteric questions along the way.

Leah Glaser helped with the initial draft of this manuscript that first summer.

Dennis Sloan answered countless questions – some of them stupid and many of them more than once – and shared his vast wealth of knowledge about the power program.

Brit Storey, Senior Historian at Reclamation, helped see this process along and provided unwavering support throughout.

Deborah Linke, current manager of Reclamation's Power Resources Office, enthusiastically picked up the project toward the end and saw it through to completion, including patiently reading and rereading drafts.

Edwina Travers graciously volunteered to be the first outside reader/guinea pig and provided timely and intelligent comments on the final manuscript.

Claire Saunders, and all the rest of my internal and external editors and reviewers, provided invaluable assistance throughout the various drafts of this manuscript.

Charles Brown, Illustrator in Reclamation's Visual Presentations Group, took my words and laid them out beautifully, giving the manuscript shape, structure, and visual aids. He then patiently dealt with all my "what happens if" questions, as well as repeated requests to move this photo, change that caption, add this section.

Many others assisted, including: a variety of summer historians who dealt with small and often esoteric research requests; Arlo Allen; Judy Austin; Jay Brigham who graciously provided a draft of his manuscript prior to its publication; Donald Bryce; Kathy Carlson; Felix Cook; Donna Daughety; Department of Energy, everyone who found me photos when I needed them; Ed Friedman; Lonnie Gourley; Cynthia Gray; Sheri Harral; James Hess; Joan Howard and the National Archives, Rocky Mountain Region; Bob Johnson; Chris Kenney; Tom Lincoln; Bruce Lonnecker; Bert Milano and the Hydroelectric Research and Technical Services Group; Christine Pfaff; Bob Quint; Tom Rawlings; Reclamation's library staff in Denver; Reclamation's regional photographers (Kelly Conner, Walter Justus, Jerry Leggate, Andy Pernick, Mark Volkoff, David Walsh); Reclamation's regional power people; Reclamation's public affairs people (Mark Andersen, Diana Cross, Kara Lamb, Jeff McCracken, Wilbert Moore, Rod Ottenbreit, Elaine Simonson, Peter Soeth, Craig Sprankle, Bob Walsh, Barry Wirth, Jone Wright); Ginger Reeve; Marty Reuss, David Treadway, and public affairs people at the Corps of Engineers; Rusty Schuster; Jay Seitz; Margaret Sibley, Elizabeth Harrison, Gayle Shanahan, and Millie Slagle, for finding the funding to complete the project; Joe Simonds; the Technical Communications Group; the Visual Presentations Group; and Roy Wingate.

Many people patiently answered my often repeated questions as I tried to understand all of this. Many of these same people also listened attentively as I talked and then corrected me when I went astray.

Others sent me documents and publications to feed my suddenly strange fascination with electricity, hydropower projects, and utility issues.

Various family, friends, and colleagues – both in and outside Reclamation – have been supportive as I've seen this project to its fruition.

Lastly, special thanks to the Bureau of Reclamation's Centennial Committee for supporting publication of this manuscript.

In any project like this, one is likely to overlook a few people when acknowledging aid and assistance. In the event I have forgotten anyone, my most sincere apologies. Even if I didn't include your name, your help was invaluable to the final product.

Toni Rae Linenberger
February of 2002

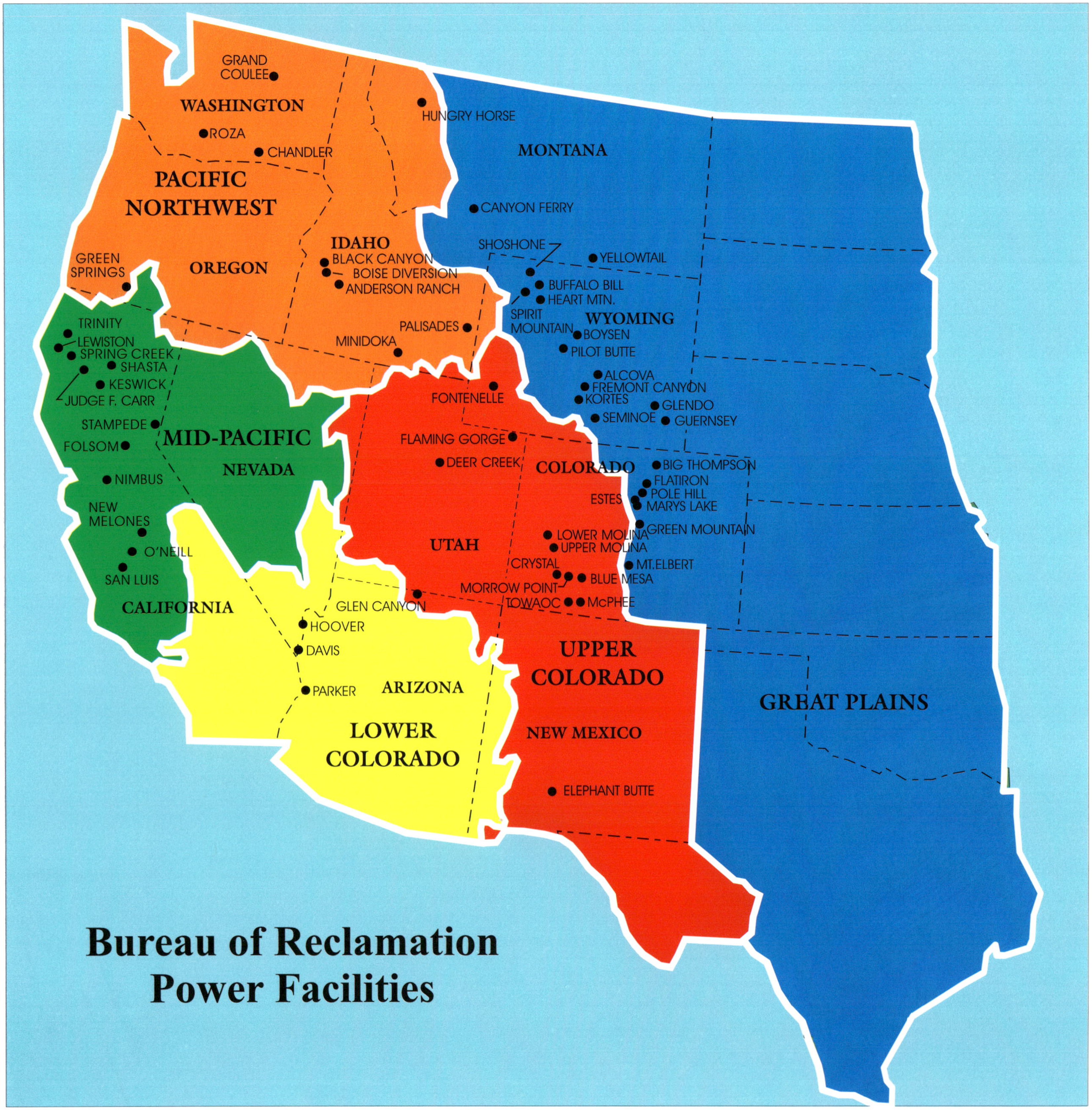

Bureau of Reclamation Power Facilities

HYDROELECTRIC POWERPLANTS WHICH THE BUREAU OF RECLAMATION OWNS AND OPERATES FISCAL YEAR 2000

REGION	PROJECT	PLANT	STATE LOCATION	RIVER	INITIAL DATE IN SERVICE	NUMBER OF UNITS	INSTALLED CAPACITY (kW)	GROSS GENERATION (kWh)
PN	BOISE	ANDERSON RANCH	IDAHO	SO. FORK, BOISE	12-50	2	40,000	151,444,000
		BLACK CANYON	IDAHO	PAYETTE	12-25	2	10,200	70,882,000
		BOISE RIVER DIV.	IDAHO	BOISE	5-12	3	* 1,500	0
	COLUMBIA BASIN	GRAND COULEE	WASHINGTON	COLUMBIA	3-41	33	6,809,000	22,960,057,750
	HUNGRY HORSE	HUNGRY HORSE	MONTANA	SO. FORK, FLATHEAD	10-52	4	428,000	1,019,921,000
	MINIDOKA	MINIDOKA	IDAHO	SNAKE	5-09	4	27,700	167,176,000
	PALISADES	PALISADES	IDAHO	SO. FORK, SNAKE	2-57	4	176,564	759,751,000
	ROGUE RIVER BASIN	GREEN SPRINGS	OREGON	TRANS. MTN. DIV.	5-60	1	17,290	72,840,000
	YAKIMA	CHANDLER	WASHINGTON	YAKIMA	2-56	2	12,000	57,915,000
		ROZA	WASHINGTON	YAKIMA	8-58	1	12,937	79,674,000
TOTAL						56	7,535,191	25,339,660,750
MP	CENTRAL VALLEY	JUDGE F. CARR	CALIFORNIA	TUNNEL, LEWISTON	5-63	2	154,400	589,781,000
		FOLSOM	CALIFORNIA	AMERICAN	6-55	3	198,720	585,273,000
		KESWICK	CALIFORNIA	SACRAMENTO	10-49	3	117,000	479,278,900
		NEW MELONES	CALIFORNIA	STANISLAUS	6-79	2	300,000	484,498,000
		NIMBUS	CALIFORNIA	AMERICAN	5-55	2	13,500	67,608,400
		O'NEILL	CALIFORNIA	SAN LUIS CREEK	11-67	6	25,200	5,428,476
		SAN LUIS	CALIFORNIA	SAN LUIS CREEK	3-68	8	** 202,000	571,745,307
		SHASTA	CALIFORNIA	SACRAMENTO	6-44	7	629,000	2,086,423,100
		SPRING CREEK	CALIFORNIA	TUNNEL, CLEAR CREEK	1-64	2	180,000	760,369,000
		TRINITY	CALIFORNIA	TRINITY	2-64	2	140,000	671,615,600
		LEWISTON	CALIFORNIA	TRINITY	2-64	1	350	2,546,050
	WASHOE	STAMPEDE	CALIFORNIA	LITTLE TRUCKEE	9-87	2	3,650	12,398,810
TOTAL						40	1,963,820	6,316,965,643
LC	BOULDER CANYON	HOOVER	ARIZ/NEV	COLORADO	9-36	19	2,078,800	5,313,796,000
	PARKER-DAVIS	DAVIS	ARIZONA	COLORADO	1-51	5	240,000	1,281,548,000
		PARKER	CALIFORNIA	COLORADO	12-42	4	120,000	538,281,669
TOTAL						28	2,438,800	7,133,625,669
UC	COLLBRAN	LOWER MOLINA	COLORADO	PIPELINE	12-62	1	4,860	14,360,800
		UPPER MOLINA	COLORADO	PIPELINE	12-62	1	8,640	24,533,650
	COLO. RIVER STORAGE	BLUE MESA	COLORADO	GUNNISON	9-67	2	86,400	281,018,000
		CRYSTAL	COLORADO	GUNNISON	6-78	1	28,000	187,463,000
		FLAMING GORGE	UTAH	GREEN	11-63	3	151,950	496,822,000
		GLEN CANYON	ARIZONA	COLORADO	9-64	8	1,296,000	4,609,522,000
		MORROW POINT	COLORADO	GUNNISON	12-70	2	173,334	359,854,000
	PROVO RIVER	DEER CREEK	UTAH	PROVO	2-58	2	4,950	26,074,777
	RIO GRANDE	ELEPHANT BUTTE	NEW MEXICO	RIO GRANDE	11-40	3	27,945	92,187,950
	SEEDSKADEE	FONTENELLE	WYOMING	GREEN	5-68	1	10,000	63,682,000
	DOLORES	MCPHEE	COLORADO	DOLORES	12-92	1	1,283	7,274,738
		TOWAOC	COLORADO	TOWAOC CANAL	5-93	1	11,495	24,163,158
TOTAL						26	1,804,857	6,186,956,073
GP	COLO.-BIG THOMPSON	BIG THOMPSON	COLORADO	TRANS. MTN. DIV.	4-59	1	4,500	9,467,000
		ESTES	COLORADO	TRANS. MTN. DIV.	9-50	3	45,000	98,641,000
		FLATIRON	COLORADO	TRANS. MTN. DIV.	1-54	3	94,500	200,521,000
		GREEN MOUNTAIN	COLORADO	BLUE	5-43	2	26,000	47,377,000
		MARYS LAKE	COLORADO	TRANS. MTN. DIV.	5-51	1	8,100	39,283,000
		POLE HILL	COLORADO	TRANS. MTN. DIV.	1-54	1	38,200	162,165,000
	FRYINGPAN-ARKANSAS	MT. ELBERT	COLORADO	TRANS. MTN. DIV.	6-81	2	200,000	376,478,760
	KENDRICK	ALCOVA	WYOMING	NORTH PLATTE	7-55	2	36,000	126,578,000
		SEMINOE	WYOMING	NORTH PLATTE	8-39	3	45,000	119,974,000
	NORTH PLATTE	GUERNSEY	WYOMING	NORTH PLATTE	7-27	2	6,400	23,968,000
	PICK-SLOAN MO. BASIN	BOYSEN	WYOMING	WIND	8-52	2	15,000	63,193,000
		CANYON FERRY	MONTANA	MISSOURI	12-53	3	50,000	312,538,000
		FREMONT CANYON	WYOMING	NORTH PLATTE	12-60	2	66,800	285,713,000
		GLENDO	WYOMING	NORTH PLATTE	12-58	2	38,000	91,818,000
		KORTES	WYOMING	NORTH PLATTE	6-50	3	36,000	119,478,000
		YELLOWTAIL	MONTANA	BIG HORN	8-66	4	250,000	727,590,000
	SHOSHONE	BUFFALO BILL	WYOMING	SHOSHONE	7-92	3	18,000	79,331,000
		HEART MOUNTAIN	WYOMING	SHOSHONE	12-48	1	5,000	15,188,000
		PILOT BUTTE	WYOMING	WIND	1-25	2	1,600	3,072,000
		SHOSHONE	WYOMING	SHOSHONE	6-92	1	3,000	22,824,000
		SPIRIT MOUNTAIN	WYOMING	SHOSHONE	10-94	1	4,500	16,929,000
TOTAL						44	991,600	2,942,126,760

Grand Total Number Of Plants: 58
Grand Total Number Of Units: 194
Grand Total Installed Capacity: 14,734,268 kW
Grand Total Gross Generation: 47,919,334,895 kWh

* BOISE RIVER DIVERSION POWERPLANT UNDERGOING REHABILITATION
** FEDERAL SHARE OF 424,000 kW INSTALLED CAPACITY PLANT OPERATED BY THE STATE OF CALIFORNIA
NAVAJO STEAM PLANT: RECLAMATION SHARE IS 546,750 kW INSTALLED CAPACITY WITH 2000 GENERATION OF 4,783,141,959 kWh

Marys Lake Powerplant. Colorado-Big Thompson Project, Colorado. Date: Unknown. Photograph courtesy of Northern Colorado Water Conservancy District.

ac	alternating current
AEC	Atomic Energy Commission
APA	Alaska Power Administration
APS	Arizona Public Service
AVA	Arkansas Valley Authority
B.C.	British Columbia
BCP	Boulder Canyon Project
BIA	Bureau of Indian Affairs
BMI	Basic Magnesium, Incorporated
Bonneville	Bonneville Power Act of 1937 or Bonneville Power Administration *aka* BPA
CAP	Central Arizona Project
Compact	Colorado River Compact
Corps	United States Army Corps of Engineers
CRDF	Colorado River Dam Fund
CRSP	Colorado River Storage Project
CUP	Central Utah Project
CVP	Central Valley Project
dc	direct current
DOE	United States Department of Energy
DOI	United States Department of the Interior *aka* Interior
DWP	Los Angeles Department of Water and Power *aka* LADWP
EA	Environmental Assessment
EIS	Environmental Impact Statement
EMC	embedded microchip
Fact Finders	Fact Finders Commission
FDR	Franklin Delano Roosevelt
FERC	Federal Energy Regulatory Commission
FPC	Federal Power Commission
FWS	United States Fish and Wildlife Service
IJC	International Joint Commission
kV	kilovolts
kW	kilowatts
Laboratory	Power Management Laboratory
M&I	municipal and industrial
MPPDA	Motion Picture Producers and Distributors of America
MVA	Missouri Valley Authority
MWD	The Metropolitan Water District of Southern California
NAERO	North American Electric Reliability Organization
NEPA	National Environmental Policy Act of 1969
NERC	North American Electric Reliability Council
NIIP	Navajo Indian Irrigation Project
NIRA	National Industrial Recovery Act of 1933
NPR	National Performance Review
OPEC	Organization of the Petroleum Exporting Countries
PG&E	Pacific Gas and Electric Company
PSMBP	Pick-Sloan Missouri Basin Program
PUHCA	Public Utility Holding Company Act of 1935
PURPA	Public Utility Regulatory Policy Act of 1978
PWA	Public Works Administration
REA	Rural Electrification Administration
Reclamation	United States Bureau of Reclamation or United States Reclamation Service or Water and Power Resources Service *aka* USBR or USRS
SCADA	Supervisory Control and Data Acquisition
SCE	Southern California Edison
Southeastern	Southeastern Power Administration *aka* SEPA
Southwestern	Southwestern Power Administration *aka* SWPA
SRP	Salt River Project
SRVWUA	Salt River Valley Water Users Association
TVA	Tennessee Valley Authority
USGS	United States Geological Survey
Western	Western Area Power Administration *aka* WAPA
WPPSS	Washington Public Power Supply System
WSCC	Western Systems Coordinating Council

Orchards and farms at Palisade. Grand Valley Project, Colorado. Date: July 17, 1908. Photographer: Walter Lubken.

H.J. Mersdorf property near Caldwell, Idaho. Boise Project, Idaho. Date: September 11, 1940. Photographer: Unknown.

A transmission line running through rugged mountains between Roosevelt Dam and the Salt River Valley. Salt River Project, Arizona. Date: February 11, 1909. Photographer: Walter J. Lubken.

CHAPTER ONE

Historians, environmentalists, and politicians generally discuss the Bureau of Reclamation (Reclamation or USBR)[1] in terms of its development of water for irrigation and domestic (municipal and industrial or M&I) uses. Historians address the positive and negative impacts of Reclamation water projects on the West over the course of history. Environmentalists lament the loss of natural habitats and the alteration of river systems due to construction of Reclamation's large, multipurpose projects. Politicians sit on either side of the fence, depending on their constituents' current views and national policy. However, there is much more to Reclamation's water development projects than just water supply.

In addition to dams, Reclamation also constructed powerplants to capture the energy produced by falling water. In turn, these powerplants helped bring electricity and modernization to vast portions of the West. Though Reclamation is better known for its role in water development in the West, Reclamation also played a very significant role in electrification of the region.

The Bureau of Reclamation is the second largest producer of hydropower in the country, surpassed only by the United States Army Corps of Engineers (Corps). Reclamation, a bureau of the United States Department of the Interior (Interior or DOI), owns and operates fifty-eight hydroelectric powerplants, comprised of 194 generating units, in eleven of the seventeen Western states in its normal area of operation. In addition, Reclamation operates three plants for other bureaus. Reclamation constructed several other plants that are now operated by other entities. Reclamation is also entitled to a portion of the power generated at the Navajo Generating Station, a coal-fired plant in Arizona. Together, Reclamation and the Corps are the United States' largest producers of hydropower.[2]

While the amount of energy Reclamation can generate depends on water supply, on average, just over forty-four million megawatt hours of energy is produced every year. An average family uses 1,000 kilawatt hours monthly.

Internationally known as one of the premier dam builders, Reclamation's hydropower program has also had a significant international presence. Among other activities, Reclamation has taken an active role in international activities, organizations, and conferences. Reclamation has offered technical assistance to a number of countries – mostly developing nations – constructing their own hydroelectric infrastructure. A study of the international impacts of Reclamation's hydropower program has yet to be undertaken and falls outside the scope of this manuscript, which focuses on the hydropower program within the United States.

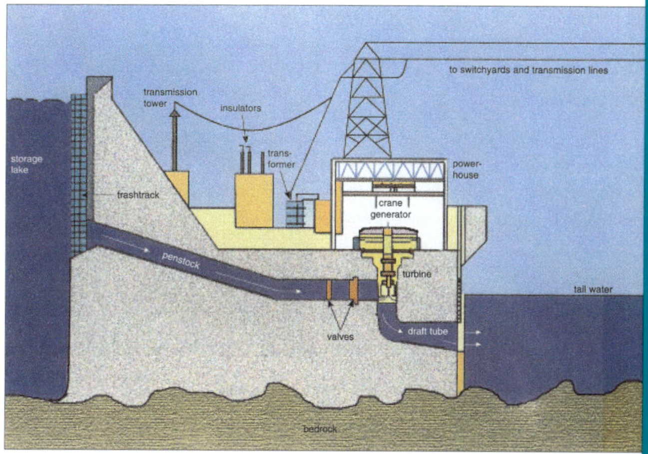

Hydroelectric generation follows a simple concept: water falling under the force of gravity turns the blades of a turbine, which is connected to a generator. Electricity generated by the spinning turbine passes through a transformer and out to transmission lines supplying factories, shops, farms, and homes.

FUNDAMENTALS OF HYDROELECTRICITY

To better understand the variety of forces and issues related to the development of Reclamation's hydropower program, a brief explanation of the fundamentals of electricity is in order. Generating electrical power converts one form of energy to another. Energy can never be created or destroyed, but it can change form.[3] Hydroelectric power production converts moving water into electrical power. Water in motion, such as in a river or waterfall, creates kinetic (moving) energy. The amount of kinetic energy depends on the speed of the water, which varies, depending on factors such as streamflow, and change in elevation (head).

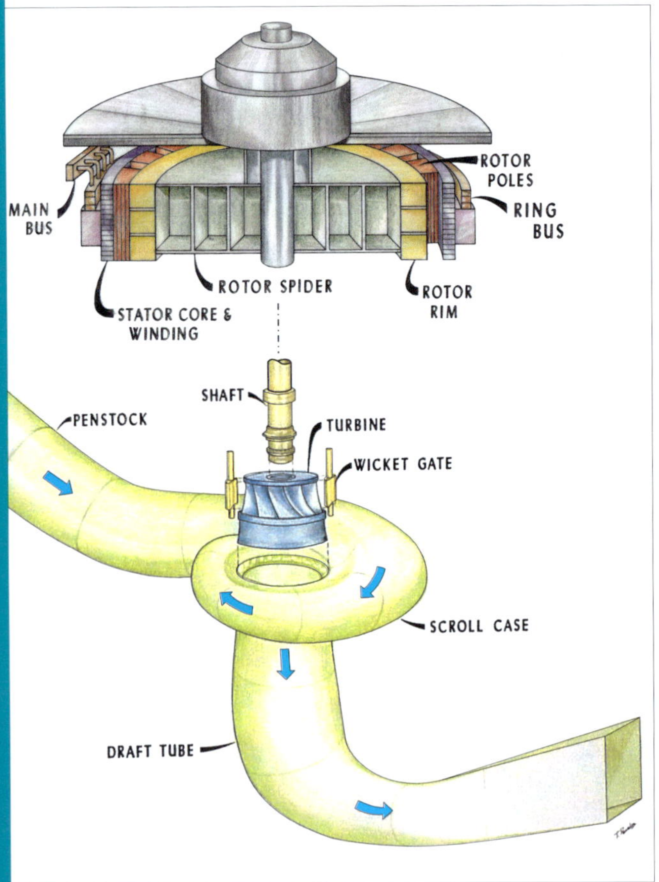

Illustration - water flow through penstock, scroll case, turbine, and draft tube.
Date: February 2002. Illustrator: Tony Rozales.

Primarily two types of dams can produce hydroelectric energy. Storage dams generate power with the release of water (usually for other purposes such as irrigation). Run-of-the-river facilities utilize the natural streamflow. Because Reclamation's water management role places a high priority on the orderly release of water for irrigation, flood control, and river regulation, its hydroelectric production occurs mainly at storage dams on multipurpose projects.

To generate hydroelectricity, water from the river or reservoir is directed into a large pipe called a penstock. Water flowing inside the penstock can then be controlled and directed to one or more generators to produce electricity (electrical energy).

All generating units have four main parts: a turbine (the operative part being the runner), a rotor, a shaft to connect the turbine to the rotor, and a stator. The falling water in the penstock is directed into the turbine, forcing the turbine to turn at a very fast speed. Since a shaft connects the turbine and the rotor, the rotor rotates at the same speed as the turbine. The outside edge of the rotor is comprised of very strong electromagnets (also known as field windings), made by wrapping large copper wires around a steel part called a pole body. These magnets are connected to provide alternating north and south poles around the rotor.

The stator of the generator is also made of steel with many coils of small copper wire inside. There are always many more stator coils than there are rotor magnets. By putting the stator around and very close to the rotor, the energy of the water pressure – which is turning both the turbine and the rotor – combined with strong electromagnets will induce (or create) voltage in the stator coils. By connecting the stator coils in certain patterns, the voltage induced can be controlled. In addition, by changing the strength of the magnets on the rotor and the amount of water flowing in the penstock the amount of electricity (energy) being generated can be regulated.

To get the electricity out of the generator, the copper coils in the stator are connected to large copper bars known as buses; these buses are then connected to a transformer. The transformers convert the electricity to a very high voltage that makes it easier to send over long distances, for use by everyone who needs it. The power of moving water (kinetic energy) has now been converted to electricity.

Once electricity is generated, it needs to be delivered or transmitted to where it can be used. Since water sources are often located long distances away from electricity customers, long-distance transmission is an important aspect of the hydroelectric power system. Large amounts of electricity cannot be economically stored, and unused electricity dissipates. Stored energy, like that of a battery, is potential energy that can be converted into electrical energy when needed. However, the process of generating electricity and storing it in batteries for later use is presently economically infeasible. However, the water used to produce electricity can be stored.

There is much more to the development of hydroelectric power than the mere conversion of one form of energy to another. The history of the development of hydropower in the West is comprised of a series of human choices often made within heated and controversial political, social, and economic forums, and among other related issues. Hydropower production by the Bureau of Reclamation encompasses issues of control, democratic ideals, *laissez-faire* capitalism and competition, economic development, equality of opportunity, and, finally, quality of life.

Rotating field rotor for the generator for Unit N-5 at Hoover Dam being erected. Field poles are ready for testing. Boulder Canyon Project, Arizona, Nevada. Date: June 24, 1947. Photographer: Ruper Spearman.

Footnotes for Chapter 1: Introduction

[1] Secretary of the Interior Ethan Allen Hitchcock originally established the United States Reclamation Service (Reclamation or USRS) under the Director of the United States Geological Survey (USGS). In 1907, the USRS became a separate bureau within the Department of the Interior. In June of 1923, Secretary of the Interior Hubert Work renamed the USRS the Bureau of Reclamation. In November of 1979, Secretary of the Interior Cecil Andrus changed the bureau's name again, this time to the Water and Power Resources Service. In May of 1981, Secretary of the Interior James Watt changed the name back to the Bureau of Reclamation. In this study, for purposes of consistency, the organization will be referred to as Reclamation or by its respective acronyms USRS or USBR. See later discussions on Reclamation's early history for additional information.

[2] United States Department of the Interior, Bureau of Reclamation, *Future Generations: A New Era of Power, Performance, and Progress* (1996).

[3] Energy exists in two forms – kinetic and potential. Potential energy is energy that is stored in the elevation of an object or fluid. Kinetic energy is energy of motion. An object flying through the air has kinetic energy. Potential energy can be converted to kinetic energy. The energy of a ball on a table is converted from potential energy to kinetic energy by pushing it off the table. Energy can never be created or destroyed but it can change form. <www.usoe.k12.ut.us/curr/science/core/8thgrd/integrated/Potentialkinetic.html> accessed September 8, 1998.

Generator in powerplant. Salt River Project, Arizona. Date: November 21, 1907. Photographer: J.M.M.

CHAPTER TWO

Although much of the initial electrical development and invention took place prior to the turn of the twentieth century, Reclamation's hydropower program can be better understood when placed into the larger framework of the development of electrical technology. In the twenty years before Congress created Reclamation, innovators and inventers like Nikola Tesla and O.H. Ensign laid the groundwork which made the success of Reclamation's hydropower program possible.

When electrical power generation and transmission entered their early developmental stage, Federal involvement in public utilities and other major industries was rare. At the turn of the twentieth century, most Americans still considered electricity a commodity for private enterprise, rather than a government service. As electrical usage increased, electric utilities began to develop into large-scale economic enterprises. This new conglomerate style of organization followed a pattern established by the railroad, petroleum, steel, and telegraph industries after the Civil War. The large companies taking over the technology of electrical power staked out service areas and formed natural monopolies that altered the Nation's economy. These conglomerates controlled electrical production from generation through transmission and distribution. These monopolies also inhibited competition within a service area, as larger companies were loath to invest in an area already controlled by another utility.[4]

Twelve-hundred kW generator, Minidoka Dam. Minidoka Project, Idaho. Date: May 3, 1909. Photographer: R.A.W.

Similarly, privately owned companies initially motivated and determined the evolution of electrical technology. As the electrical industry became increasingly consolidated, so did the tools of invention. Larger corporations absorbed the independent inventors responsible for the initial technological developments. In many cases, these independent inventors had been inventing largely for their own curiosity and interest – they were not driven by economic pressures. When they were absorbed by larger corporations, inventors became innovators, as companies called for improved productivity and efficiency from existing technologies, in contrast to the earlier focus on new invention. Along with "innovation," "planning" and "coordination" became important buzzwords in the developing electric industry.

The late nineteenth century was a period of rapid and independent innovation in electrical technology. Although electrical service was generally limited to the northeastern and midwestern parts of the country, America led the world in electrical system development. Scientists experimented with electricity as early as the 1700s, but the invention of arc lights in the 1840s stimulated great interest and spurred rapid technological development.

Control panel, Black Canyon Dam Powerplant. Boise Project, Idaho. Date: Unknown. Photographer: Unknown.

Throughout the 1880s, Thomas Edison issued several licenses to various companies to use his discoveries, including, but not limited to, incandescent lights and direct current (dc) electrical equipment. The companies using Edison's patents eventually merged to become the General Electric Company. Although about fourteen companies competed in the electrical arena at the time, the Westinghouse Electric and Manufacturing Company, established by George Westinghouse in 1884, was Edison's strongest competitor.

By 1892, Edison and Westinghouse emerged as the national industry leaders. To avoid costly and time-consuming litigation, Edison and Westinghouse pooled their patents and sued other potential competitors to keep them out of the industry. As the electric utility industry expanded and consolidated, problems inherent in the integration of several different systems with a variety of equipment types became evident.[5]

Alternating Current versus Direct Current

Because of corporate competition and the subsequent push for advancement in technological innovation, the early electrical industry did not standardize its equipment and had few compatible transmission networks. Throughout the 1880s, the most popular debate in the industry concerned the best, most efficient, and economical mode of electrical transmission – direct or alternating current (ac). This controversy helped to further fragment the industry and prevent any widespread coordination of the systems.

Both dc and ac have their own unique properties – a fact early electrical inventors and distributors quickly learned. One of the most important early lessons was that the higher the force (voltage) of the electricity being transmitted, the less energy lost during transmission. Because current is reduced in the transmission of electricity at higher voltages, less energy is lost during transmission.[6] This became especially important as the distance increased between the point of generation and the point of use.

Direct current flows in one direction. This constant flow of current does not allow for the use of a transformer to either increase or decrease the voltage of the electric current. Historically, this has made direct current harder to transmit, as it must normally be transmitted at the same voltage at which it was generated. The inability to increase the voltage of dc made it economically and practically inefficient for long-distance transmission. Today, modern technology of inversion and rectification equipment can be applied to interconnect dc transmission lines to ac systems.

Despite its early history of inefficiencies in transmission, Thomas Edison and the Edison Company advocated the wide-spread adoption of dc as the industry standard. Conversely, the Edison Company's competitor, George Westinghouse, advocated the industry-wide adoption of ac as the standard.

Alternating current reverses direction (or oscillates) in the circuit at regular intervals. The number of times ac oscillates is known as " frequency." Frequency (measured in "cycles per second" or "hertz") can be varied, although the objective is to maintain a constant value. The accepted standard frequency for electrical distribution in the United States is 60 hertz; in Europe, the accepted standard is 50 hertz. Alternating current allows the use of transformers, located at either end of the transmission system, to step voltage up or down. By increasing the voltage being transmitted, less energy is lost during transmission, making the system more economical and efficient.

NIKOLA TESLA AND HIS AC MOTOR

The ac system, however, could not be practical until an efficient, compatible electric motor was developed. In reality, the ac system did not lack an efficient electric motor; rather, builders of ac systems lacked the knowledge that one existed. As early as 1878-79, Elihu Thomson built a simple, though inefficient, electric motor, while Lucien Gaulard and Lucien Gibbs produced the first alternating current transformer, a necessary component for increasing and decreasing voltages in power transmission. Westinghouse quickly bought the American rights to Gaulard and Gibbs' patents to further his crusade to standardize and promote ac systems.

In 1882, while in Budapest, Serbian-born Nikola Tesla conceived a simply constructed ac motor. Tesla's motor was an induction motor and was light-years ahead of the other ac motors in existence at the time. The only problem was that much of the design and construction of the motor was not documented. Because Tesla lacked the necessary funding to construct a prototype, a practical ac motor remained the unrealized dream of a young inventor.

Over the next several years, Tesla focused his energies on various other projects. He took a job with the telegraph office and worked his way up to engineer. In the fall of 1882, he moved to Paris and began a job with Edison's telephone subsidiary. One of the goals foremost in Tesla's mind when he took the new job was selling

Interior view of powerhouse, showing the exciters in place and ready for operation. Strawberry Valley Project, Utah. Date: December 26, 1906. Photographer: Walter Lubken.

Water wheel at Boise, Idaho. Boise Project, Idaho. Date: April 25, 1901. Photographer: A.L.F.

the officers of the Continental Edison Company on the benefits of alternating current. His hopes were dashed, however, when the officers told him of Edison's aversion to the subject. While on a trip to Strasbourg, Germany, to cure the ills of Edison powerplants, Tesla constructed his first ac induction motor in his spare time. The prototype proved highly successful; however, Tesla still lacked the funds to do anything more than experiment with his model. He also lacked the foresight to begin patenting his inventions.

A few years later, Tesla emigrated to the United States and began working directly for Edison. However, in part because of Edison's aversion to alternating current, the two inventors did not see eye-to-eye on many things, and Tesla soon set out to form his own independent company. After at least one failed attempt at starting his own company and several odd jobs, Tesla finally secured the funding to develop his ac induction motor and founded the Tesla Electric Company. With the financial support of the Western Union Telegraph Company, among others, Tesla went about making his inventions a reality. Between 1887 and 1891, Tesla applied for and received a total of forty patents, all related in some way to his ac motor and the resulting ac system he devised.[7]

Two 300 kW turbo generators, power station Number 1. North Dakota Pumping: Williston Unit, North Dakota. Date: May 2, 1908. Photographer: L.P.S.

Tesla's patent applications occurred at an opportune moment. His work provided the missing piece George Westinghouse had been waiting for to make his ac system more efficient in order to compete with Edison's. Westinghouse quickly capitalized on the inventor's ideas by buying the rights to all of Tesla's patents and offering him a job as a consultant, adapting the technology to Westinghouse's system. Tesla took the job.

One of the first problems Tesla encountered while constructing the new ac system was the difference in frequency. Westinghouse's system used 133-cycle current, while Tesla designed his induction motor to use 60-cycle current. After months of disagreement with the engineers and several failed experiments, Tesla prevailed and Westinghouse's system became a 60-cycle system.[8]

The adoption of Tesla's motor further inflamed what became known as the "War of the Currents." Westinghouse's now-successful ac system could compete with, and even surpass, Edison's dc system. Where Edison previously had a monopoly on electrical distribution – without any alternatives he gained support by default – he now had to fight to maintain his dominance in the field. The Menlo Park – Edison's New Jersey laboratory – propaganda machine soon started advocating widespread adoption of dc current as the standard. To further his point, Edison characterized ac as more dangerous than dc.

Determined to win the war of the currents, Edison waged a no-holds-barred campaign to convince the public of the dangers of ac.[9] In what he considered the final *coup de grace*,

Edison, with the help of Samuel Insull and former lab assistant Harold Brown, quietly bought the rights to three of Tesla's patents without revealing their intended use. Patents in hand, Brown paid a visit to Sing Sing Prison in New York. Soon after, prison officials announced they would begin executing prisoners on death row using a newly- crafted electric chair which utilized ac. The first electrocution of a condemned murderer took place August 6, 1890, against George Westinghouse's opposition.

Westinghouse endured Edison's underhanded tactics and continued to cite facts and figures to reinforce the safety of ac. Eventually, the advantages of ac, including the increased capacity for long-distance transmission, prevailed and the industry chose ac as its standard.[10]

DEVELOPMENT OF LONG-DISTANCE TRANSMISSION

Advances in the development of ac energy enabled the development of longer transmission lines. Initially, urban centers, with their high profit potential and relatively short transmission requirements, were the power companies' primary market. Without the ability to transport electrical energy over long distances, electrical generation could not reach rural areas like farms and mines, places that are usually far removed from powerplants. The same was true of certain electrical energy sources. Hydropower could produce a large quantity of electrical energy, but depended on long-distance transmission to put it to use, since hydropower sites were often located in remote areas.

The most significant advances in hydroelectric and transmission technology developed in California during a period of corporate mergers and industry consolidation. As smaller transmission and distribution systems consolidated into larger ones, or were absorbed by corporations, the resulting systems were integrated with one another and standardized for ease of transmission. Consequently, throughout the 1890s, electrical equipment became more standardized.[11] Out of this process of standardization came the opportunity for innovation and refinement of the existing systems for maximum efficiency. One such technological refinement occurred just prior to the turn of the twentieth century.

Built from 1894 to 1897, the Santa Ana transmission line in southern California, at the time of its construction, was the longest transmission circuit in the country. It surpassed the Folsom to Sacramento line completed in 1895. The Santa Ana line was also one of the most powerful. Eighty-two miles in length, it had the capacity to operate at 33,000 volts (thirty-three kV) of electricity down the Santa Ana Canyon into the San Bernardino Valley. This ambitious project was the brainchild of engineer Orville Hiram Ensign. (Ensign later went to work for Reclamation.) One of the keys to the line's success was his

High-voltage terminal (bushing) on a newly installed, 230 kV power transformer at the Mt. Elbert Pumped-Storage Powerplant. The bushing is being prepared for a low-voltage impulse test to assess the transformer winding. Fryingpan-Arkansas Project, Colorado. Date: 1977. Photographer: Unknown.

Insulators

development of the so-called "Redlands type" insulator.

Insulators support the conductors – usually a bare wire – and keep them away from the transmission tower or pole; this prevents the current from going to ground and causing a short circuit. Insulators are made of a material, such as porcelain or glass, that does not conduct electricity. Prior to Ensign's glazed porcelain insulator design, insulators that could withstand voltages higher than ten kV were rare. Porcelain was much stronger than glass, and the flared-out edges (referred to as a "petticoat") further increased the insulation value of the insulator. The arrangement of wires and insulators on the poles was also important to the line's success.[12]

Electrical engineers considered a variety of factors when determining the best tower design. These included wind, sleet, terrain, temperature, and the probability of accidents, such as breaking cables or towers. At the time, the structures most adaptable to these types of conditions and purposes for transmission structures were pyramid-shaped windmill towers. Designers of high-voltage electrical lines often turned to windmill manufacturers to design transmission towers. These new tower designs further aided efforts to standardize long-distance transmission, making long-distance transmission of electric power considerably more feasible and affordable. The ability to achieve long-distance transmission of electrical power greatly influenced the use of hydroelectricity. With new tower designs and the ability to transmit power over long distances, it was no longer necessary to be close to water sources to use hydroelectric power.[13]

Outside of the historic Folsom Powerhouse.
Date: October 1998.
Photographer: T.R. Linenberger.

FOLSOM DAM AND POWERHOUSE

Beginning in the 1860s, Horatio Gates Livermore dreamed of constructing a dam on the American River in Folsom, California, and turning the small mining town into an industrial center. Livermore realized the necessity of a consistent water supply to make his dream a reality. To achieve his goal, Livermore did several key things. He acquired land, bought shares in the local water and mining company (Natoma Water and Mining Company), and began advocating for the construction of Folsom State Prison – a second state correctional facility to accommodate the overflow from San Quentin. By 1874, Livermore and his sons controlled the Natoma Company and owned more than 9,000 acres of land surrounding the town. The same year, construction began on Folsom State Prison.[14] The Natoma Company deeded the land for the prison to the State of California with the stipulation that convict labor would be used to construct a dam for the mining and water company.

By the time construction began on Livermore's long-planned dam, it was the late 1880s, and his attention had turned from creating an industrial center to the burgeoning hydroelectric industry: Livermore dreamed of supplying hydroelectric power to Sacramento and other points. That dream did not prove completely infeasible. Workers completed the original Folsom Dam and power canal in 1893. The elder Livermore died that same year; however, his two sons, Charles E. and Horatio P., continued their father's crusade. All that remained was harnessing the power of the river. To these ends, work soon began on a sawmill and a powerhouse.

For many years, the elder Livermore tried to find a way to float logs down the river to Folsom. As a result, the completed dam included log booms in the forebay. Logs could then be floated through the power canal to a mill pond and sawmill operated by the American River Land and Lumber Company, an affiliated enterprise of the Natoma Company. Water from the river was used to generate electricity. When completed, the sawmill became the first electrically powered mill in the United States.[15]

Convinced the future of the Nation rested in electric power, the two brothers set about constructing a powerplant with the intention of supplying power for electric trams in Sacramento, twenty-two miles away. Knowing they needed outside assistance, the Livermores sought financial assistance and support from several companies, including the General Electric Company, which contributed funding to the project. Completion of the system in 1895, "marked a new achievement in long-distance transmission at high voltage."[16] The appurtenant transmission line carried three-phase alternating electric current from Folsom to Sacramento, making it the longest transmission line in the world, until the Santa Ana line took that title in 1897.

In 1903, financial difficulties forced the Livermores to sell the Folsom Powerhouse, as well as their other interests, to the California Gas and Electric Company, predecessor of the Pacific Gas and Electric Company (PG&E). The sale ended more than fifty years of involvement in Folsom's development by the Livermore family.

The historic powerhouse continued to operate until 1952, when the Corps demolished the old Folsom Dam to make way for a newer dam and powerplant upstream.[17] In 1958, PG&E turned the powerhouse complex over to the California State Parks System for preservation and interpretation. The powerhouse remains a listed State and National Historic Landmark, a National Civil Engineering Landmark, and is on the National Register of Historic Places.[18]

TURBINES

The development of high-voltage transmission lines and associated features of a hydroelectric system were not the only factors contributing to the feasibility of long-distance electrical transmission. As historian Thomas P. Hughes observed, "the highly developed state of water-turbine technology, at the time high voltage transmission became feasible, greatly facilitated the use of hydroelectricity."[19] The wide variety of turbine designs decreased in the 1880s as the industry

Outside of Folsom Dam. Folsom, California. Central Valley Project: Folsom Unit, American River Division, California. Date: October 1998. Photographer: T.R. Linenberger.

Exciter inside historic Folsom Dam Powerhouse. Date: October 1998. Photographer: T.R. Linenberger.

Cutaway of Reaction (Francis) Turbine.

Cutaway of Impulse (Pelton) Turbine.

adopted specific individual turbine design and construction.[20] Scientists first developed power turbines and generators, legacies of the ancient water wheel, in early nineteenth-century France. Water-driven mills played a prominent role in thousands of rural and village communities around the world. For example, mill systems in Lowell, Massachusetts, harnessed the power of an entire river.

The amount of power generated at a facility depends the pressure of the water where it enters the turbine housing. The force and speed of water often depend on a concept defined as "head." Engineers adapted turbines to a wide range of heads, depending on the height of the storage dam or natural waterfall. Turbines produce power according to water height, pressure, and velocity, or any combination of the three. An engineer has to determine the type and setting for a particular system. For example, the Pelton turbine, or the impulse wheel, is the only turbine operating with just one force (velocity), and system designers often use it in high head and low flow developments.

Turbines fall into two general classes – the impulse type and the reaction type. An example of the impulse type is a Pelton turbine (wheel), while both Propeller and Francis turbines are examples of the reaction type. An impulse turbine converts the static pressure of water into the kinetic energy of a high-pressure water-jet. The buckets of the impulse turbine runner have a twin cup design. The high-velocity water-jet hits the bucket at the interface of the two cups. The jet and runner of an impulse type turbine are surrounded by air. The reaction type turbine develops power primarily from the difference in water pressure between the entrance and exit of the runner, and from the velocity of the water through the runners. In contrast to the impulse type, water completely fills the runner and water passages. Francis turbines have a relatively large number of shrouded blades or buckets. Propeller turbine runners, as the name implies, resemble the propeller of a ship. The runner may have three to six unshrouded blades that may be fixed or adjustable. A propeller unit with adjustable blades is commonly referred to as a Kaplan turbine.[21]

On hydroelectric projects, the turbine wheel is generally smaller than on water power systems, such as mill operations. Water power systems use the force of water to directly power machinery. Reclamation commonly uses Francis and Pelton type turbines. Both the Francis and Pelton designs are based on European research and engineering with American craftsmanship and manufacturing.

For both turbines, design concerns included wheel efficiency, generator compatibility, increased wheel capacity, and adaptation of turbines to high heads. Not surprisingly, engineers involved in the construction of the massive hydroelectric facilities at Niagara Falls in the 1890s resolved many of these issues. Minor design modifications continued throughout the early part of the last century, but the basic design remained unchanged.

"Inside the Turbine." Workmen repairing turbine scroll case, Grand Coulee Dam. Columbia Basin Project, Washington. Date: Unknown. Artist: Fletcher Martin.

In the late 1960s, the Bureau of Reclamation embarked on a program to present its accomplishments to the public through the medium of art. Forty of America's most prominent artists were commissioned to visit Reclamation's water resource development sites and record their impressions in whatever medium and style they wished. The artwork in this document came from this program.

Footnotes for Chapter 2: The Early Development of Electrical Technology

[4] David Nye, *American Technological Sublime* (Cambridge, Massachusetts: MIT Press, 1994), 139-141.

[5] David Nye, *Consuming Power* (Cambridge, Massachusetts: MIT Press, 1998), 124-128.

[6] During transmission, electricity dissipates due to the resistance between the transmission line and the electricity traveling across it. This dissipation of energy is called loss.

[7] The ac system quickly came to be preferred over the dc system because of enhanced ac motor and generator performance. Alternating current easily transformed voltage levels and, thus, reduced losses, ensuring economical long-distance transmission of energy. In addition, ac could be easily interrupted, as opposed to designing more complex and expensive breakers to interrupt dc currents.

[8] In her biography on Tesla, Margaret Cheney argued that Westinghouse's adoption of a 60-cycle system, necessary to efficiently run Tesla's motor, set the later standard for alternating current within the industry. Margaret Cheney, *Tesla: Man Out of Time* (New York: Barnes and Noble Books, 1993).

[9] In addition to issuing pamphlets and making speeches against ac, Edison resorted to the electrocution of neighborhood pets and animals to demonstrate the dangers of the system. He then called this "being Westinghoused."

[10] Cheney, 23-45.

[11] For a comprehensive description of the development of electrical transmission and hydroelectric power, see Mark T. Swanson, *Santa Ana River Hydroelectric System* (National Park Service, Historic American Engineering Record, No. CA-130), 1992.

[12] William A. Myers, *Iron Men and Copper Wires* (Glendale, California: Trans-Anglo Books, 1984), 38-39; for considerable detail on Orville Ensign and the development of the Santa Ana transmission line, see Swanson, 79-87.

[13] D.R. Scholes, "Fundamental Considerations Governing the Design of Transmission-line Structures," *Proceedings of the American Institute of Electrical Engineers* (June 29, 1908): 1011-1018; P.M. Downing, "Foundations for Transmission Line Towers and Tower Erection," *Proceedings of the American Institute of Electrical Engineers* (June 20, 1915): 1227-1233.

[14] Initial construction of the prison was completed in 1878. The prison was unusual for its time because of the electric light plant built on the grounds. The two-dynamo plant lit both the interior and the exterior of the prison. Later, a second prison powerhouse was constructed using convict labor. The prison powerhouse contained six Laffell turbine water wheels. The generated electricity powered irrigation pumps for the prison ranch and twelve hoisting engines in the stone quarry and prison stone yard.

[15] The majority of mill equipment ran on power captured from falling water. Usually, a water wheel used the energy of water passing over it or, in some cases, under it, to power an interconnected system of gear wheels and pulleys, which then ran any necessary machinery.

[16] Quoted in Par Environmental Services, Inc., *Cultural Resources Investigation for the American River Bridge Crossing Project, City of Folsom, Sacramento County, California* (Sacramento, California: Par Environmental Services, Inc., 1991), 28.

[17] While originally built by the Corps as part of the Central Valley Project (CVP), Reclamation operates Folsom Dam and Powerplant in conjunction with the rest of the project.

[18] Par Environmental Services, 26-29.

[19] Thomas P. Hughes, *Networks of Power: Electrification in Western Society, 1880-1930* (Baltimore, Maryland: The Johns Hopkins University Press, 1983), 263.

[20] Louis H. Klotz, "Water Power: Its Promises and Problems" (Durham, New Hampshire: University of New Hampshire Center for Industrial and Institutional Development, February, 1980), 32; Louis Hunter, *A History of Industrial Power in the United States, 1780-1930 Vol. 1: Water Power* (Charlottesville, Virginia: University of Virginia Press, 1979), 292-342.

[21] Hughes, *Networks of Power*, 263; William P. Creager and Joel D. Justin, *Hydroelectric Handbook* (New York: John Wiley and Sons, Inc. 2nd edition, 1950), 811-815; Edwin T. Layton, Jr., "Scientific Technology, 1845-1900: The Hydraulic Turbine and the Origins of American Industrial Research," *Technology and Culture* 20 (January 1979): 88; Hunter, *Water Power*, 392-394.

Potatoes grown on the Charles Hogenson Ranch, near Burley, Idaho. Minidoka Project, Idaho. Date: October 22, 1909. Photographer: Unknown.

CHAPTER THREE

Settlement of the lands west of the Mississippi River required fortitude, perseverance, and a lot of imagination, particularly when devising ways to irrigate the arid lands of a region known as the "Great American Desert." Water, often plentiful in the winter when portions of the region received extensive snowfall, disappeared in late summer, just when crops needed it the most. New Western residents, accustomed to the plentiful water supplies of the East, became dismayed at the lack of natural water and began searching for alternatives.

A specimen of rare Crown Saguaro Cactus on the other side of Fish Creek Hill, Arizona. Salt River Project, Arizona. Date: 1907. Photographer: Walter J. Lubken

"The desert." Salt River Project, Arizona. Date: 1911. Photographer: Walter J. Lubken.

For the most part, alternatives meant irrigation works of one form or another. Unfortunately for individual residents, irrigation works required engineering expertise and extensive construction work as well as considerable funds, things most residents lacked. Sometimes, small communities joined forces to build irrigation systems for the entire group; however, these efforts often failed due to a lack of knowledge, cooperation, and/or funds. A few endeavors of this type succeeded, typically when something other than just the common desire for water held the residents together.

The most notable group of settlers to initiate mass irrigation works was the Mormons,[22] who settled in Utah in 1847. Bound together by religion and a desire to succeed, the Mormons successfully irrigated large portions of Utah. Horace Greeley, editor of the *New York Tribune*, also initiated a reasonably successful irrigation colony. Greeley, Colorado, started out as a Utopian colony in 1870, but its residents

The damming of City Creek by Mormon Pioneers, July 23, 1847.

eventually rejected farming, and it became a more traditional city.[23] Inspired by these successes, or at least partial successes, other communities tried to emulate their methods. Most quickly exhausted their funds and turned to the Federal government for help.

THE UNITED STATES ARMY CORPS OF ENGINEERS

The first official Federal influences on national waterways occurred in the eastern United States and were primarily the responsibility of the United States Army Corps of Engineers. The origin of the Corps dates back to 1775, when Congress established a Continental Army with provisions for a Chief Engineer. After the successful conclusion of the Revolutionary War, members of the Corps were discharged from the service in 1783, along with most of the Continental Army. In 1802, the United States permanently reestablished a separate Corps of Engineers and founded the United States Military Academy at West Point, New York, under the supervision of the Corps.

In 1819, Secretary of War John C. Calhoun published a report on the importance of waterways for national defense, commerce, and economic development.[24] The same year, Major Stephen H. Long began his expedition up the Missouri River and began the Corps' involvement in Western exploration. It was on this expedition that Long characterized the arid area west of the Mississippi River as the "Great American Desert."

In 1824, the Corps' mission expanded to include the construction of permanent civil works when Congress passed an act to improve the navigation of the Ohio and Mississippi Rivers. At the same time, the General Survey Act authorized the use of Army engineers to survey road and canal routes. In 1879, the Corps continued its involvement in United States water development with the creation of the Mississippi River Commission to execute a comprehensive flood control and navigation plan on the Lower Mississippi River. On March 3, 1899, Congress enacted a law, "Structures on Navigable Waters," which required approval by both Congress and the Chief of Engineers before building a structure on any navigable body of water.

Wool train on road near Carson, Nevada. Newlands Project, Nevada. Date: April 28, 1901. Photographer: Frederick H. Newell?

The Crabtree family cabin at the entrance of Hell's Hip Pocket in Brown's Basin between Four Peaks and Salt River. Salt River Project, Arizona. Date: January 1908. Photographer: Walter J. Lubken.

The Corps entered into the Federal hydroelectric business in 1925, with completion of Wilson Dam[25] at Muscle Shoals on the Tennessee River.[26] Although the Corps eventually built structures throughout the United States, in its early years, it primarily worked on developing waterways in the East. By the middle of the twentieth century, the Corps emerged as the largest constructor and operator of Federal hydropower facilities in the United States.

Early Settlement Efforts in the West

Similar to its way of dealing with the power industry, the Federal government had long supported a policy of nonintervention when it came to the West and its settlement. For example, in 1866 Congress passed "An Act Granting the Right-of-Way to Ditch and Canal Owners Over the Public Lands . . ."[27] which effectively left all irrigation development to "local customs, laws, and decisions of courts." The Desert Land Act followed just over ten years later on March 3, 1877. This act allowed both individuals and companies to claim up to 640 acres of arid land for $1.25 an acre, after supplying water to reclaim the land for three years.

The Carey Act, passed in August of 1894, allowed public-land states containing desert lands to select up to a million acres of land to be irrigated, reclaimed, and occupied; in this manner, the Federal government made the individual states into trustees for settlers. The Carey Act required individuals to reclaim the land in parcels limited to no more than 160 acres. The Carey Act also provided for state-constructed hydraulic facilities – including dams, reservoirs, canals, and laterals. States did have the option to contract with private companies to actually construct hydraulic works. In turn, the act required companies, or the state, to submit their plans for irrigation works to the Secretary of the Interior for approval prior to the start of construction.

While the Federal government was beginning to recognize the importance of at least some governmental influence in the development of western irrigation, settlers found state efforts lacking.[28]

On May 24, 1869, Major John Wesley Powell and the Powell Geographic Expedition set out from Green River, Wyoming, on the first expedition down the Colorado River. Powell began his

Old abandoned camp at Goldfield, Arizona. Salt River Project, Arizona. Date: March 26, 1907. Photographer: Walter J. Lubken.

John Wesley Powell. Date: Unknown.
Photographer: Unknown.

expedition on the Green River, a tributary of the Colorado. The expedition arrived three months and six days later, minus three men and a boat, at Grand Wash Cliffs in Arizona.[29] By the end of his journey, Powell had realized that all the water in the West could not "make the desert bloom." Powell detailed his findings and his own conclusions in his landmark government sponsored report, *Report on the Lands of the Arid Region of the United States*, published in 1878. Powell's report provided the impetus for a national irrigation movement.

THE NATIONAL IRRIGATION MOVEMENT

National irrigation crusades, led by irrigation proponents such as William Ellsworth Smythe and George Maxwell, began soon after publication of Powell's report, but, for the most part, Westerners' pleas fell on deaf ears in Washington. The Federal government did not yet accept the idea that irrigation, or "reclamation" as the effort had come to be known, of the western United States should be a national effort. The term "reclamation" referred to an effort, by either an individual or group, to "subjugate" or "reclaim" arid lands for human settlement and use. Nevertheless, in 1888, Congress appropriated funds and authorized the United States Geological Survey (USGS) to investigate the lands and rivers of the West and determine their irrigability.[30]

As national attention turned towards irrigation, local irrigation efforts continued unabated, as did efforts to convince Congress and the President of the validity of a national irrigation movement. Irrigation failures and petitions for national assistance finally paid off for Westerners when Theodore Roosevelt succeeded to the office of President, after the assassination of William McKinley in September of 1901. Roosevelt understood the importance of irrigation, and Westerners suddenly found a sympathetic audience for their pleas. In addition to being a Westerner at heart, Roosevelt was a staunch conservationist and progressive. He believed that the good of the many outweighed the good of the few and that natural resources ought to be managed in such a way as to ensure sustained exploitation. Planning ways to federally mandate his ideas, Roosevelt set about creating the United States Forest Service and

President Theodore Roosevelt speaking at the dedication of Roosevelt Dam.
Salt River Project, Arizona.
Date: March 1911.
Photographer: Walter J. Lubken.

orchestrating the passage of the Federal Reclamation Act.

The primary motivation behind the Reclamation Act was a desire to conserve the West's water and make the Great American Desert hospitable to settlement. To an early nineteenth-century American, conservation of water meant not letting any of it go to waste. To them, water that was not being used was being wasted. To these nineteenth-century Americans, the primary purpose of water was to irrigate a field, meaning any water not being used for irrigation was going to waste. If all the water in riverbeds were diverted to irrigate crops, the productivity of the Nation would go up; and many believed "rain would follow the plow."

Irrigation advocates speculated that if settlers plowed the usable farmland in the West, the freshly turned soil would attract additional rain to the area. Furthermore, the additional rain would help "make the desert bloom" and, by extension, make the West look more like the East, which was the vision of ideal settlement at the time.

Pioneer Day Parade in Ogden. Float of some Ogden county products. Ogden River Project, Utah. Date: July 24, 1934. Photographer: Tom Mead.

In the minds of many Americans, irrigation equaled civilization. In addition to increasing the productivity of the Nation as a whole, it was widely believed that by turning the current population into farmers the Reclamation Act would civilize and settle a lawless region.[31] Later, electricity became as important to people's ideas of civilization as water.

Ruins of old Arizona State Prison, Yuma, Arizona. Yuma Project, Arizona. Date: April 26, 1932. Photographer: George A. Beyer.

THE RECLAMATION ACT

Encouraged by President Roosevelt's exuberance for western development, Representative Francis Newlands, from Nevada; George Maxwell, leader of the National Irrigation Association; and Frederick Newell, of the United States Geological Survey, Division of Hydrography, rewrote one of the many irrigation acts previously presented to Congress. The resulting bill, the Reclamation Act of 1902, passed Congress with relative ease, and on June 17, 1902, President Roosevelt signed the act into law. The Reclamation Act was also known as the Newlands Act after the Nevada Senator (he was elected to the Senate in 1903) who promoted its adoption.

Reclamation's Chief Engineer (Arthur Powell Davis), Director (Frederick Haynes Newell), Mr. H.N. Savage, and Mr. L.C. Hill at Arrowrock. Boise Project, Idaho. Date: August 16, 1911. Photographer: Walter J. Lubken.

The Reclamation Act authorized the Secretary of the Interior to locate and construct irrigation works in sixteen of the seventeen Western states and territories;[32] a subsequent act in 1906 added Texas, rounding out the seventeen. In July, Secretary of the

Charles D. Walcott, Director, USGS (1894-1907); Director, USRS (1902-1907).

Senator Francis G. Newlands, Senator Franklin Mondell, and L.H. Taylor, project engineer for the state of Nevada, standing on newly completed headworks on Carson River, just after speaking at the opening of the new irrigation project. Newlands Project, Nevada. Date: June 17, 1905. Photographer: Walter J. Lubken.

Interior Ethan Allen Hitchcock established the United States Reclamation Service within the USGS; Charles D. Walcott, Director of USGS, became the new bureau's first Director and Frederick Newell its first Chief Engineer.

Though many different Reclamation laws had been proposed, this was the first that had the full support of the Chief Executive. After spending much of his adult life in the West, Theodore Roosevelt felt that conservation of our natural resources, especially water and forests, was important for the survival of the Nation. "While I had lived in the West I had come to realize the vital need of irrigation to the country," Roosevelt later wrote.[33] It was this zest for conservation that led Roosevelt to support the Reclamation Act, as well as to create the Forest Service. He strongly encouraged his two new "pet projects" to work in conjunction with one another in their efforts to preserve the country's natural resources.

The Reclamation Act provided assistance to Westerners for construction of irrigation works, though not necessarily in the form anticipated. The act placed the Secretary of the Interior almost wholly in charge by making him responsible for withdrawal of lands from the public domain, selection of suitable sites for study and construction, and, if necessary, contracting with private entities for construction of project facilities. The act further limited homesteads on Reclamation projects to 160 acres per person, allowing a husband and wife 320 acres total, and required residence on irrigated lands.

With Presidential and Secretarial backing, the engineers of the new Reclamation Service began an active program of investigation and construction in the West. One of the first projects undertaken was the Truckee-Carson Project in Nevada, authorized March 14, 1903. Actual construction began in August of 1903, and was completed in 1915. In 1919, the Truckee-Carson Project was renamed the Newlands Project in honor of the Senator who helped create the Reclamation Service. Including Truckee-Carson, the Secretary of the Interior approved a total of five projects on March 14, 1903 – the Salt River (Arizona), Uncompahgre (Colorado), North Platte (Nebraska and Wyoming), and Milk River (Montana) projects. As construction began on the Newlands Project, Reclamation's engineers continued to look for other sites to irrigate. Authorizations continued steadily for the next several years for a total of twenty-five projects during the first five years of the program.[34]

Derby Diversion Dam dedication. Newlands Project, Nevada. Date: June 17, 1905. Photographer: Unknown.

THE RECLAMATION FUND

The Reclamation Act set aside monies received from the sale of public lands in the West (minus the five percent already allocated the states for education) for a special fund in the U.S. Treasury, known as the Reclamation Fund. Money was roughly allocated to the state where the funds originated to examine, survey, construct, and maintain irrigation works for the storage, diversion, and development of waters to reclaim arid and semiarid lands. This act originally excluded Texas because it contained no public lands for sale.[35]

Settlers on project lands were required to repay project construction costs into the Reclamation Fund. According to the act, repayment would be made within ten years, and the Secretary of the Interior was authorized to determine the amount and repayment schedule.

In 1906, the mission of the USRS expanded. The Reclamation Act of 1906 (June 12) authorized the USRS to investigate and construct works in Texas. The Town Site and Power Development Act (April 16, 1906) added development of hydropower resources to Reclamation's overall purpose, provided that the development and lease of power not needed for project purposes did not impair the efficiency of the irrigation project. This act further authorized the Secretary of the Interior to sell surplus power generated at Federal Reclamation projects, giving customer priority to municipal purposes, such as water pumping and street lighting. The receipts from power sales returned to the Reclamation Fund as credits to repay construction costs on the associated project.[36]

Generally, the early projects produced power during the summer, while making releases for irrigation by running the water through the turbines to produce electricity before delivering water to irrigators. When water releases ceased, so did hydroelectric production. Gradually, especially after reservoirs filled to capacity, Reclamation began releasing water on a year-round basis which, in turn, increased opportunities for power production. However, power generation continued to depend upon available water; extra water was not released for power purposes. Even today, power development and production generally remain secondary to water storage, delivery, and flood control, even though in many cases hydropower energy production has become the "paying partner of irrigation."

Elephant Butte Dam looking east.
Rio Grande Project, Texas, New Mexico.
Date: October 29, 1915.
Photographer: Unknown.

REPAYMENT OF RECLAMATION PROJECTS

In 1914, the Reclamation Extension Act further modified the Reclamation Fund. The 1914 Act stipulated that as a condition of receiving lands settlers on each project were required to repay the Reclamation Fund, on a strictly scheduled basis, any construction costs incurred as a result of the project. The repayment schedule was no longer to be determined by the Secretary of the Interior; instead, repayment would begin five years after completion of all construction activities. Depending on terms of the repayment contract, once settlers paid off the project or construction activities were completed, they assumed the management and operation of the facilities.

Secretary Hubert Work and Commissioner Elwood Mead at Stony Gorge Damsite. Orland Project, California. Date: March 26, 1927. Photographer: Unknown.

More importantly, the 1914 Act made the Reclamation Fund subject to Congressional appropriations as opposed to the sale of public lands, a drastic change from the original intent of the Reclamation Act. This made Reclamation dependent on Congressional inclination when it came to construction funds for new and existing projects. However, it also freed Reclamation from the financial constraints imposed by the original Reclamation Act and allowed for the large-scale construction activities that followed in the 1920s and 1930s.

By 1923, twenty-one years after its inception, Reclamation had initiated construction on more than twenty-five irrigation projects in the western United States. Unfortunately, its efforts did not always prove entirely successful for those who chose to settle on the projects. Despite the fact that the government charged no interest on its investment, and later increased the repayment period by ten years, many residents abandoned irrigated farms because they were unable to repay construction loans. Those who remained on the projects often fell hopelessly behind in their payments to the U.S. Treasury.

THE FACT FINDERS COMMISSION

In September of 1923, at the request of Secretary of the Interior Hubert Work, a group of prominent citizens agreed to serve on the Fact Finders Commission to study the ailing Bureau of Reclamation. Secretary Work had renamed the United States Reclamation Service the previous June. The commission was originally comprised of Thomas E. Campbell, a recent governor of Arizona; John A. Widtsoe, an agricultural scientist and former university president; Oscar E. Bradfute, president of the American Farm Bureau; James R. Garfield, former Secretary of the Interior; and Clyde C. Dawson, in place of President of the United States Chamber of Commerce Julius H. Barnes. Elwood Mead joined the commission in January of 1924.[37]

The Fact Finders took their role very seriously and spent the next seven months investigating Reclamation. Based on their investigations, which

Rise of groundwater near Fallon, Nevada. Newlands Project, Nevada. Date: Unknown. Photographer: Unknown.

included a conference in Salt Lake City, Utah, attended by representatives from all but two of the existing Reclamation projects, the Fact Finders made their recommendations to Congress. Congress, in turn, considered the recommendations and incorporated them into legislation known as the Fact Finders' Act (Second Deficiency Appropriation Act for 1924), signed December 5, 1924.

The Fact Finders' Act mandated several things designed to make Reclamation projects more successful. This act called for careful investigations of projects prior to authorization, the screening of prospective settlers, construction charges that reflected soil types and income derived from the land, and turning over operation and maintenance of irrigation systems to the settlers. This act also provided that after water users took over the operation and maintenance of a project, revenues derived from the operation of project powerplants, leasing of project grazing and farm lands, and the sale or use of town sites be credited to the construction charges of the project.

Another direct impact of the act was the execution of new repayment contracts on twenty-one projects.[38] The Fact Finders recognized what was already happening on projects like Minidoka and Salt River Project (SRP); namely, that power revenues were being used to help defray repayment costs. Realizing this philosophy could be applied to other projects with a power component, they made provisions for power revenues to be credited to project repayment. By doing this, the Fact Finders firmly entrenched the role of power production in Reclamation's activities, and acknowledged a solution to the high costs of project repayment that a few projects had already adopted.

EARLY RECLAMATION POWERPLANTS

Even prior to passage of the Town Site and Power Development Act of 1906, Reclamation engineers discovered that electricity was a necessary component of many irrigation works. In addition to aiding the delivery of water, primarily through electric pumps, electricity also proved beneficial in construction activities. Engineers used electric power to process materials and construct engineering works; a powerplant could

Mr. Leslie McClellan, Chief Electrical Engineer, and Dr. Elwood Mead, Comissioner, about to board a plane for Chicago. Date: January 28, 1932. Photographer: George A. Beyer

Interior of powerhouse at government construction camp No. 1, view showing air compressor for operating drills in excavation of Tunnels No. 1 and 2. The steam engine drove the generator furnishing electric power for tramming and lighting in Tunnel No. 1, operating concrete plant and carpenter shop, and lighting camp. Grand Valley Project, Colorado. Date: August 19, 1913. Photographer: Henry T. Cowling.

later be used to pump water through the irrigation system, if necessary. Plants powered cableways, trams, concrete plants, draglines, hoists, giant shovels, sawmills, and lights that facilitated round-the-clock operation.[39]

Even in the early stages of project investigations, hydropower development played a role in project planning. In some areas, local electric companies existed and Reclamation could utilize their power resources – either by tying into or extending an existing transmission line, building a new transmission line, or, if luck would have it, utilizing an already existing line that reached the construction site. If a local power company did not already deliver power to an area, Reclamation might find a new power company willing to extend service to the construction site. For the most part, both of these options worked reasonably well; however, in a few instances, circumstances forced Reclamation to construct its own plant. These initial plants were never intended to do anything more than facilitate construction of a given project.

At some of these early plants, including one on the Uncompahgre Project in Colorado, Reclamation used steam-powered, instead of hydro, plants constructed on site to meet the electrical needs of the project. After completion of construction activities, Reclamation dismantled the plants since they were no longer needed. The need for steam plants on certain projects did not end when Reclamation officially entered the hydropower business. As late as 1913, Reclamation constructed a steam plant on the Grand Valley Project to provide electric power for tramming and lighting in Tunnel No. 1, operating a concrete plant and carpenter shop, and lighting the construction camp.

THE STRAWBERRY VALLEY PROJECT

The Upper Spanish Fork Powerplant, on the Strawberry Valley Project in Utah, was an early hydropower plant designed originally for project construction purposes. Initial investigations of the project began in 1903 and included a hydropower component.

A powerplant was completed and began generating power in 1908. Power generated at the plant was used at the Strawberry Tunnel and Dam during construction activities. The tunnel was completed in 1912 and the dam in 1913, though the entire project was not completed until 1922. After completion of the project, power generated at the initial powerplant and two subsequent powerplants was used for irrigation and municipal purposes. The Strawberry Water Users Association continues to operate the plants.

THE MINIDOKA PROJECT

Even prior to official acknowledgment of commercial hydropower development on irrigation projects, engineers and survey crews seeking suitable irrigation sites throughout the western states kept in mind the potential for hydropower development on a project. Reclamation engineers began investigating one such project, the Minidoka Project in Idaho, as early as 1903. The initial plan for site development included plans for a powerplant.[40] Minidoka later became one of the first two commercial Reclamation powerplants, along with Roosevelt on the Salt River Project; both plants went on line in 1909.

Reclamation originally built the Minidoka Powerplant to provide electricity for pumping irrigation water to the project's bench lands on the south side of the Snake River. To add an additional 50,000 acres of irrigable lands to the project that could not be fed by gravity irrigation, Reclamation opted to add the powerplant after completing the dam in 1907. Construction of the powerplant proved valuable to the local economy.

Reclamation soon discovered that the irrigation project only required electricity during the summer months to run irrigation pumps. However, the Snake River at the Minidoka Powerplant provided a relatively constant supply of water year-round, leaving surplus water to generate electricity during the spring and winter months. Reclamation and the local residents capitalized on the surplus.

Beginning in 1910, Reclamation sold surplus energy to local residents for home heating, industry, and other uses. In many cases, the residents formed

View of the control panel in the Strawberry Valley powerplant. Strawberry Valley Project, Utah. Date: March 18, 1965. Photographer: Stan Rasmussen.

Rupert Electrical High School. Minidoka Project, Idaho. Date: August 1914. Photographer: Henry T. Cowling.

Domestic science room in Rupert Electrical High School. Minidoka Project, Idaho. Date: February 1914. Photographer: Dibble.

their own cooperative utility companies and built their own transmission lines. Reclamation then sold the energy to the utility cooperatives, which, in turn, resold it to their members. Electrical production on the project greatly enhanced the local quality of life. Thanks, in large part, to Reclamation development, in 1914, the town of Rupert, Idaho, attracted national attention by building the first electrically heated high school building in the United States.[41]

Theodore Roosevelt Powerplant. Salt River Project, Arizona. Date: Unknown. Photographer: Unknown.

THE SALT RIVER PROJECT

The Minidoka Project was not the only early Reclamation project where hydropower development affected the local community in positive ways. The Salt River Project in Arizona also allowed for significant development of the local community, both through its irrigation works and its development of hydropower.

Beginning in the 1860s, settlers dug canals to irrigate lands north and south of the Salt River in central Arizona. These cultivated lands yielded highly profitable crops; however, a scarce and poorly controlled water flow hampered further agricultural expansion.[42] Valley farmers responded to passage of the 1902 Reclamation Act by forming the Salt River Valley Water Users Association (SRVWUA) to lobby for a Federal water storage project. On March 12, 1903, Secretary of the Interior Ethan Allen Hitchcock selected SRP to be one of the first five Reclamation projects in the country. In 1904, the government began to build Roosevelt Dam on the Salt River, sixty miles northeast of Phoenix. Studies estimated that Roosevelt could probably generate enough hydropower to pump underground water to irrigate 50,000 additional acres.

Generator shaft, Theodore Roosevelt Powerplant. Salt River Project, Arizona. Date: Unknown. Photographer: Unknown.

Roosevelt Dam workers first constructed a wood-burning steam unit to supply power for hoisting and concrete aggregate processing and for machine and carpentry shops. Soon after, they altered the design of a second temporary powerplant. The original plans called for locating construction generators within the reservoir area. Instead, workers erected a temporary hydropower plant at Roosevelt Dam that provided the 1,200 to 1,300 horsepower needed to run machinery, operate the cement mill, and provide lighting for constructing the dam. Reclamation installed three power canal units to convey water to a seven-foot-diameter, steel-lined penstock tunnel that delivered water to the powerplant's generating units.[43]

As at Minidoka, government officials soon realized the high potential for the commercial sale of hydroelectric energy to the Salt River Valley. Although the government intended that the electricity generated at Roosevelt Dam would be used for pumping groundwater, they also realized it had commercial possibilities. The 1906 Town Site and Power Development Act allowed SRP to use the receipts from these energy sales as credit toward repaying the cost of building their reclamation project. Immediately following Congress's decision,

Reclamation extended Roosevelt's penstock tunnel and built a permanent plant to house the power canal units. Three additional units received water from a ten-foot-diameter penstock running through the dam face. By 1916, SRP boasted six turbine generating units.[44]

The concept of marketing and selling electricity over long transmission lines opened rural farms and remote mining communities as new economic markets. Long-distance transmission capabilities brought electricity to previously unreachable areas. As long-distance transmission became more widespread, the need for usable materials for construction of transmission lines increased. Copper, an ideal material for conveying electricity, was an obvious choice and the copper industry quickly expanded, aided in part by the electrical industry. As a matter of fact, when the railroad arrived in central Arizona in the 1880s, the large copper deposits became one of Arizona territory's most important exports.[45]

The commercial opportunities for hydropower in the sparsely populated and mountainous western states region were numerous. Electrical machines optimized production and reduced labor costs in mines and on irrigation projects, and electrical pumping maximized water supply and delivery. Industries that could benefit from the use of hydropower included agriculture and irrigation, textile mills, mining, and the railroads.[46]

The Federal government and SRP farmers proposed selling hydroelectric energy for revenue, as long as the power production did not decrease the water supply at Roosevelt Reservoir.[47] In 1907, Secretary of the Interior James R. Garfield agreed to the future sale of 1,500 kilowatts (kW) of generation capacity at Roosevelt Dam to the Phoenix-based Pacific Gas and Electric Company (later Central Arizona Light and Power Company) for distribution to the city of Phoenix and to various Salt River Valley businesses.[48]

The distribution of power from Roosevelt Dam also fell on Reclamation's shoulders. Engineers based the design for the transmission line from Roosevelt Dam to Phoenix on the standards of the emerging electrical industry. These included the pioneering techniques their colleagues had used to design the hydroelectric systems at Niagara Falls, New York, and in California.[49] To direct the development of Arizona's system, the government hired Orville Hiram Ensign, who had worked for the Redlands Electric Light and Power Company in southern California when it built the first long-distance, three-phase, alternating current electrical system in the United States. His first assignment with Reclamation took him to the Salt River Valley, where he applied the knowledge and expertise he gained in California to a similar climate and terrain.

In central Arizona, the terrain, wind and arid weather conditions, and high voltage required the strength and stability of steel towers rather than wood or steel poles. A large portion (thirty miles) of the transmission line, from Roosevelt Powerplant to Phoenix,

Erecting a transmission line in the desert. Salt River Project, Arizona. Date: March 20, 1908. Photographer: Walter J. Lubken.

Valve inside Roosevelt Powerhouse.
Date: Unknown.
Photographer: Walter J. Lubken.

Manufacturing a wooden ball in Powell, Wyoming, for closing a forty-two inch penstock at Buffalo Bill Dam. Shoshone Project, Wyoming.
Date: November 1932.
Photographer: Unknown.

traveled through a desert mountain region where soil conditions were often loose and rocky. After Ensign assessed the topography and determined certain specifications, Reclamation officials solicited bids from local agents of various national companies to manufacture the towers.[50]

The United States Wind Engine and Pump Company of Batavia, Illinois, won the contract for its galvanized steel windmill towers. While the windmill company supplied the towers, Reclamation engineers specified parts, materials, and measurements. Although adaptations to the towers and line expansion continued through the 1920s, the completion of the line marked the beginning of a manageable and successful partnership between water storage and power production in the government.[51]

Bringing Roosevelt Powerplant up to its potential generating capacity was an important concern because of the competitive power market, rising electrical rates, and the project's need for revenue to pay both operation and maintenance costs and repayment costs. However, Roosevelt's primary function of irrigation restricted its hydroelectric potential. Without regulating the flow of the Salt River, Roosevelt could only generate power when irrigation needs required water releases. In 1917, after the SRVWUA took over operation of the project, it set about repaying the government for its hefty cost. Over the next decade, with the aid of a new president and an experienced water and power development engineer, the water users constructed two more dams, Mormon Flat and Horse Mesa, which served to expand the project's hydroelectric capacity without significantly affecting the water supply for irrigation needs. This allowed SRVWUA to earn tremendous revenues selling hydroelectric energy to area mines and the city of Phoenix.[52]

Hydroelectric power development on the Boise, Huntley, and Shoshone Projects followed on the heels of hydroelectric development on the Minidoka and Salt River Projects. Each one provided surplus energy that Reclamation sold to local industries, businesses, and residents. These projects also resulted in Reclamation expanding its activities into the construction of transmission lines and delivery of electrical power.

Footnotes for Chapter 3: The Origins of the Bureau of Reclamation and Its Power Program

[22] Mormon is a commonly accepted term denoting a member of the Church of Jesus Christ of Latter-Day Saints (also LDS), founded by Joseph Smith in 1830.

[23] Donald J. Pisani, *From the Family Farm to Agribusiness: The Irrigation Crusade in California and the West, 1850-1931* (Berkeley: University of California Press, 1984), 51-53, 120-121.

[24] United States Army Corps of Engineers, *The History of the U.S. Army Corps of Engineers* (Alexandria, Virginia, 1998), 37.

[25] The Corps began construction on Wilson Dam in 1918 to answer a pressing need for nitrates used to make explosives; Congress authorized construction of the dam and two nitrate plants in 1916. Construction activities were suspended at the end of World War I. The Corps finally completed the structure in 1925. The nitrate plants were then used to make fertilizer instead of explosives, though some munitions research was done at the plants. In 1933, Congress turned the dam over to the Tennessee Valley Authority (TVA) as part of the TVA Act. John H. Kyle, *The Building of TVA: An Illustrated History* (Baton Rouge: Louisiana State University Press, 1958), 58-61.

[26] Corps, *History of the U.S. Army Corps of Engineers*, 54.

[27] The Act of July 26, 1866, is sometimes referred to as the Mining Act of 1866. Section 9, which deals specifically with water, is sometimes referred to as the Water Rights Act of 1866.

[28] Donald J. Pisani, *To Reclaim a Divided West: Water, Law, and Public Policy, 1848-1902* (Albuquerque: University of New Mexico Press, 1992), 88-89, 252-253; Alfred R. Golzé, *Reclamation in the United States* (Caldwell, Idaho: Caxton Printers, Ltd., 1961), 15-19.

[29] Marc Reisner, *Cadillac Desert: The American West and Its Disappearing Water* (New York: Viking, 1986; reprint, New York: Viking Penguin Books, 1993), 26-34.

[30] Major Powell, then head of the USGS, led the first of three separate investigations (the other two in 1890 and 1891) to further investigate and map the lands of the arid region.

[31] Articles to this effect appeared in such diverse and unlikely places as, among others, *Harper's Weekly*, *The Cosmopolitan*, *The American Monthly Review of Reviews*, and *National Geographic Magazine*, as well as in regional and national newspapers. All the articles shared common themes on the widespread benefits of irrigation in the West.

[32] Arizona, California, Colorado, Idaho, Kansas, Montana, Nebraska, Nevada, New Mexico, North Dakota, Oklahoma, Oregon, South Dakota, Utah, Washington, and Wyoming. Oklahoma joined the union in 1907, while New Mexico and Arizona remained territories until 1912.

[33] Theodore Roosevelt, *An Autobiography* (New York: Macmillan, 1913), 428.

[34] Michael Robinson, *Water for the West: The Bureau of Reclamation, 1902-1977* (Chicago: Public Works Historical Society, 1979), 19-20.

[35] United States Department of the Interior, Bureau of Reclamation, *Federal Reclamation and Related Laws Annotated*, Volume I of IV, through 1942, eds. Richard K. Pelz (Washington, D.C.: U.S. Government Printing Office, 1972), 31-89, 186-200.

[36] Bill Beck, "Arc Lighting and the Birth of Public Power," *Public Power* (July-August 1998); the *1906 Town Site Act*, *Statutes at Large* 34, section 5, 116 (1906); *Federal Reclamation and Related Laws Annotated*, 109-113; Norris Hundley, jr., *Water and the West: The Colorado River Compact and the Politics of Water in the American West* (Berkeley: University of California Press, 1975), 119.

[37] Elwood Mead, a university professor and world-renowned expert on irrigation, joined the commission the following January, because he was out of the country when the commission first convened. Mead became Commissioner of Reclamation in 1924 at the recommendation of the committee and served as Commissioner until his death in 1936.

[38] *Federal Reclamation and Related Laws Annotated*, 316-328; Brian Q. Cannon, "'We Are Now Entering a New Era': Federal Reclamation and the Fact Finding Commission of 1923-1924," *Pacific Historical Review* 66 (May 1997): 185-211.

[39] Robinson, 27-28.

[40] Department of the Interior, United States Geological Survey, *Second Annual Report of the Reclamation Service, 1902-1903* (Washington, D.C.: Government Printing Office, 1904), 258.

[41] Robinson, 27-30; Susan M. Stacy, *Legacy of Light: A History of Idaho Power Company*, (Caldwell, Idaho: Caxton Printers, 1991), 56-57; George Wharton James, *Reclaiming the Arid West: The Story of the United States Reclamation Service* (New York: Dodd, Mead, and Company, 1917), 154-155.

[42] Christine Lewis, "A History of Irrigation on the Tempe Area." (M.A. Thesis, Arizona State University, 1963).

[43] David Introcaso, *Mormon Flat Dam* (National Park Service, Historic American Engineering Record, No. AZ-14, 1989), 11-28. For a full description on the history and construction of the Roosevelt Powerplant, see Christine Pfaff, *Theodore Roosevelt Dam, Power Plant* (National Park Service, Historic American Engineering Record, No. AZ-6-A, 1996).

[44] In the *1906 Town Site Act*, *Statutes at Large* 34, section 5, 116 (1906), Congress authorized the Reclamation Service to sell surplus power not needed for irrigation purposes to municipalities.

[45] Bisbee, Jerome, Clifton, and Clarkdale were other places where copper was heavily extracted. Thomas Sheridan, *Arizona: A History* (Tucson: The University of Arizona Press, 1995), 161-186.

[46] David B. Rushmore and Eric A. Lof, *Hydro-electric Power Stations* (New York: John Wiley and Sons, 1923), 1-9.

[47] David Nye, *Electrifying America: Social Meanings of a New Technology 1880-1940* (Cambridge, Massachusetts: The MIT Press, 1990), 300.

[48] For a more thorough discussion on the project's hydroelectric development, see Introcaso, 1-20; James M. Gaylord, "Power and Pumping System of the Salt River Project, Arizona." (Phoenix: Salt River Project Archives, January 1, 1914), 6.

[49] Pfaff, *Theodore Roosevelt Dam, Power Plant*, 17.

[50] Leah S. Glaser, *The EMA Transmission Line* (National Park Service, Historic American Engineering Record, No. AZ-6-B, 1996), 9.

[51] Glaser, *The EMA Transmission Line*, 9.

[52] Introcaso, 22-33.

View showing a staged fault test in the 230 kV switchyard at Parker Dam. Parker-Davis Project, Arizona, California. **Date: February 2, 1947. Photographer: George O. Bonawit.**

CHAPTER FOUR

Today, most Americans view electricity as a necessity, but, prior to World War I, the government and private utilities had to convince people of the uses and benefits of electricity. In the decades surrounding the turn of the last century, many in the United States regarded electrical power as either a commodity or a public spectacle. Most people did not understand the technology of electricity and characterized it either as a mysterious agent or something godlike.

Those responsible for marketing electricity to the general public chose to capitalize on this lack of understanding and promoted the "magical" and surreal aspects of electricity in lighting displays and marketing campaigns. During the early decades of the twentieth century, the world's fairs provided an ideal opportunity to prove the new technology. Exhibitors took advantage of the captive audience, creating lavish spectacles for the public's enjoyment. The spectacle of the fairs extended to hydropower plants themselves.

As with other industrial symbols of the day, such as the Brooklyn Bridge, the American public found hydropower plants enormously appealing sights. Places like Niagara Falls became tourist attractions for the technology found there, not just the natural wonder. Some visitors considered the hydroelectric developments at Niagara Falls more inspiring than the Falls themselves.[53]

Eventually, the spectacle and magic of electricity gave way to practical uses. People overcame their fears as electricity became available to the general public. By World War I, the use of electricity was more widespread. Electric utilities in America evolved from small, local distribution systems to regional transmission monopolies.[54]

The rapid technological development and expansion of the electrical industry, along with its promises and economic potential, coincided with, and became entangled in, various political battles. Electricity first emerged as a political issue between public and private industry when the two disagreed over the rates for public lighting and street cars. The rise of monopolies in large industries was hardly unique to electrical utilities. Conglomerates like John D. Rockefeller's Standard Oil, Andrew Carnegie's U.S. Steel, and Western Union Telegraph Company emerged as targets of the reformers during what historians refer to as the Progressive Era, from the 1880s through the 1920s.

EARLY REGULATION EFFORTS

Historians generally consider politics during the Progressive Era a reaction to corporate capitalism and unregulated *laissez-faire* economic competition. Political parties like the Populists rose with an agenda to attack large corporations. They accused American industrial or financial magnates, such as Cornelius Vanderbilt and James J. Hill, of achieving their vast wealth by unethical means, such as cornering the stock market, exploiting the labor

Sun River Diversion Dam. Sun River Project, Montana. Date: 1920. Photographer: R.B. Dame.

force, or using suspicious political practices. These men earned the nickname "robber barons" after medieval European feudal lords who robbed travelers passing through their territory. Such types of corporate abuses prompted Congress to pass the Sherman Anti-Trust Act in 1890. The Sherman Act did not outlaw all monopolies but limited abusive strategies, such as predatory pricing.[55] Similarly, the rapid expansion of the electrical industry and its opportunities invited the same type of exploitation and accusations of corruption.

Secretary of the Interior James Garfield speaking at opening of Huntley Project before raising gates at headworks. Huntley Project, Montana. Date: June 26, 1907. Photographer: W. Currie.

Theodore Roosevelt at the dedication of Theodore Roosevelt Dam. Salt River Project, Arizona. Date: March 18, 1911. Photographer: Unknown.

The technological nature of public utilities discouraged competition. As a result, a few companies were created with natural monopolies. These monopolies could charge exorbitant rates and, thus, limit the distribution of power to only certain segments of society. Private power proponents, usually political conservatives and utility officials, believed electricity should follow the pattern of *laissez-faire* economics. Others agreed that electricity should stay in private hands but needed regulation by a third party so that private industry could profit but still charge customers reasonable rates. Many public power proponents feared this limitation would stall modernization of society in general and advocated public ownership.[56]

President Theodore Roosevelt believed in corporate regulation and the duty of the government to support competition. Roosevelt's beliefs led him to wage a public campaign against the developing monopolies. During his time in office, Roosevelt launched twenty-five anti-trust suits under the Sherman Anti-Trust Act. In an initial effort to oversee and regulate the electric industry, in 1906, Congress granted itself the authority to issue permits for potential hydroelectric dam sites. The passage of this 1906 law further strengthened the already complex directives of Federal authority over navigable rivers and streams by adding another checkpoint to those already in place – the Department of the Interior, Secretary of War, the White House, and now Congress. In 1908, President Roosevelt exerted his influence by vetoing a bill authorizing construction of a dam on the James River in Missouri, charging that "the bill granted an unconditional and monopolistic privilege without ensuring adequate public protection." Soon after, he asked Congress to consider generalized hydroelectric power regulation.[57]

While many progressive politicians and small businessmen argued about the regulation of utilities, the conservation movement of the early twentieth century lent its voice to the debate over hydroelectric production on the Nation's waterways. As a proponent of regulation, an active conservationist, and a strong supporter of Federal reclamation, Theodore Roosevelt promoted the development of hydroelectric power on public lands. Along with politician James Garfield, he considered it an inexhaustible and clean energy source. Conservationists such as Gifford Pinchot believed natural resource development should benefit as many people as possible in order to truly modernize the country, and hydroelectric power fit neatly into this belief. High-voltage transmission lines allowed people far away from project sites to enjoy electricity. They also believed that multipurpose operations needed public ownership – particularly those where the watershed extended across state lines and carried a national interest in the Federal oversight of inland waterways.

Hydropower carried the promise of better air quality and was quite inexpensive after the initial construction. As a power source, it seemed ideal for the American West, with its abundance of natural resources and sparse population.

Support for Public Power

Theodore Roosevelt's hand-picked successor for the Republican Presidential nomination, William Howard Taft, also promoted regulation. During his Presidency, Taft supported establishment of a Federal income tax, regulation of the telephone and telegraph industries, and the direct election of Senators (previously elected through the state legislatures, which were traditionally controlled by the railroad industry). In 1912, Democrat Woodrow Wilson won the Presidential election. Like his predecessors in the Executive Office, he also strongly advocated "trust-busting."

When President Wilson entered the White House, hydropower was the fastest growing area of electrical production, despite the fact that it generated only sixteen percent of the total electricity in the United States. With most of the technological developments of the electrical industry in place at this time, only political and economic issues blocked the spread of electricity. Throughout the 1920s, politicians debated over whether the private or the public sector should control the power industry on both national and local levels. The national crisis of the Great Depression presented public power proponents with a tremendous opportunity.

Well into the twentieth century, as electrical technology grew more sophisticated and could feasibly reach more people, both the industry and the public began to wonder if the new

Raising a corner transmission pole at the North Dakota Pumping Plant. Williston Project, North Dakota. Date: July 17, 1907. Photographer: L.P.A.

Meter gaging flow of Merced River in Yosemite National Park. Date: 1920. Photographer: R.B. Dame.

energy product should be treated as a necessary service, like water, rather than as a commodity guided by corporate practices. Concerned over the undue influence of private industry, Congress directed the Secretary of Agriculture to conduct a large-scale investigation into hydropower development in the United States. The resulting 1916 report revealed that eighteen private companies controlled half of the hydropower in the United States, but fifty-six percent of hydropower in the West had ties to public lands. This information indicated that Federal officials in Washington had a legitimate interest in making various decisions concerning hydropower and the economic development of the American West.

Gaging station, Tuolumne River, Hetch Hetchy, Yosemite National Park. Date: 1920. Photographer: R.B. Dame.

CONSOLIDATION OF PRIVATE UTILITIES

The rapid advancement of electrical technology from the 1890s through the 1910s created a perplexing problem by today's standards. It allowed the production of more energy than people in many parts of the West could use, or even knew how to use. Engineers perfected most of the major components of electrical technology before 1920. Between 1920 and 1930, generating units and transmission voltages increased in size. Since that time, engineers have achieved better frequency control as well as higher steam pressures and temperatures in turbines. They have also improved transmission and gained a greater understanding of circuitry and lightning protection. The most significant breakthrough of the 1920s occurred with the introduction of automated substation equipment, which was adopted for widespread use by the middle of the decade.

As the basic electrical system developed, the use of electricity increased, particularly in industrial circles. The share of the power market used by industry increased from twenty-eight percent in 1912 to more than fifty percent in the 1920s. These shifts in the utility industry paved the way for a fundamental shift in how management invested its energy and money. With a reasonably efficient and reliable system in place, managers of utilities now focused on how to maximize profit from the electric systems. Investment in innovation from individual inventors was no longer so important. At the same time, holding companies, which were outside of state jurisdiction, began to take control of the utility industry by consolidating small, private systems into large conglomerates.[58]

The sudden prevalence of holding companies within the utility industry spawned numerous political and economic debates during the 1920s. Public power proponents believed private utilities, once they became part of holding companies, thwarted the social benefits of regulation and expanded service. Despite such objections, industry consolidations continued during the 1920s. Over the next fifteen years, five holding companies controlled eighty percent of national electrical generating capacity. These companies primarily marketed to industry rather than to domestic households. Public power interests controlled the remaining twenty percent. By 1932, three holding companies controlled half of the electric utility market, and many analysts have since blamed their unsound business practices for contributing to the 1929 stock market crash. Meanwhile, Westerners began to urge development of more public hydroelectric sites that could serve new markets in their region at low cost.

FEDERAL INVOLVEMENT IN POWER GENERATION

The rise of the public power movement, advocating publicly owned utilities from local and municipal levels to the Federal level, began around 1913, with the City of San Francisco's effort to dam Hetch Hetchy Valley in Yosemite National Park. From 1913 to 1928, the Hetch Hetchy Valley, Great Falls on the Potomac, Muscle Shoals in Tennessee, Boulder Canyon on the Colorado River, and the Saint Lawrence River hydropower projects shaped public policy at the national level.

Yosemite National Park, Merced River below Washington's Column and Stoneman Bridge–Half Dome on right. Date: 1920. Photographer: R.B. Dame.

Congressional hearings debating the benefits of *laissez-faire* operation over public ownership finally culminated in passage of the first Federal Water Power Act in 1920. The act symbolized a significant victory for public power. While the act promoted and regulated private power, it also protected Federal involvement. The 1920 legislation established the Federal Power Commission (FPC), comprised of Secretaries of the Interior, War, and Agriculture, and granted the new commission authority to issue fifty-year permits for hydroelectric dams on the country's navigable waters.

Based on the Federal government's jurisdiction over interstate commerce and its ownership of various water power sites in the West, officials in Washington, D.C., could influence most of the hydroelectric development in the country.[59] The new commission also required that private parties request clearance to build on both navigable and non-navigable waterways, established a clear application and approval process, and gave public interests preference in its decisions.[60]

Establishment of the FPC cemented Federal involvement in the expansion and regulation of hydropower and other electric energy resources. Federal involvement in what originated as a largely private enterprise began with the exigencies of industrial mobilization for World War I. The need for planned industrial production highlighted the need for additional sources and supplies of electricity to maintain a high level of production. At the same time, Federal officials looked to electricity as a way to fuel a wartime economy.

Secretary of the Interior Ray Lyman Wilbur driving silver spike during construction of the railroad used to build Hoover Dam. Boulder Canyon Project, Arizona, Nevada. Date: September 17, 1930. Photographer: Unknown.

WORLD WAR I

World War I exposed immediate problems within the existing energy system. Just prior to the United States' mobilization, transportation of coal between mines and various points of consumption required one-third of the total carrying capacity of the railroads. Wartime industrial activities increased the demand for coal, while at the same time, troop mobilization engaged the national railroad system in movement of men and supplies to eastern ports for departure to Europe. The increased usage of the railroad for personnel and supply movement, and an increased demand for coal, overtaxed available resources. The Nation lacked sufficient coal cars to transport coal to where it was needed, while the addition of passenger trains to transport troops created massive traffic jams on rail lines throughout the country.

To alleviate part of the strain on the overworked transportation system, the government gave utilities top priority for delivery of coal because they used the resource more efficiently than industrial factories. Electricity, generated on a large scale in a central location, could be widely distributed to run electric machinery within factories. Government-mandated cooperation between utilities and industry, and the

Spud Special at the Payette, Idaho, railroad depot, as part of the train's tour to publicize Idaho potatoes. Photo courtesy of Idaho State Historical Society, 75-142.1g.

need to efficiently shift power from one place to another, spawned greater interconnections between systems.[61] Unintentionally, the need to relieve the transportation crisis provided another argument for advocates of public power and a push for additional hydropower development.

Necessity, in the form of an overtaxed railroad system, added pressure for centralized production of electricity. Moving the Nation's energy in the form of electricity rather than coal relieved the strain on the railroads and helped further develop the transmission system.

Even as the government advocated the central production of electricity and interconnection of utility systems, government interests turned an eye towards additional development of electrical resources. Development of hydropower resources could save precious resources like coal, and the government could take advantage of its jurisdiction over hydropower sites to involve itself in the utility business. To these ends, the government began construction of Wilson Dam at Muscle Shoals on the Tennessee River to provide power for explosives manufacturing plants. With construction of the Tennessee Valley Authority dams in the 1930s, this area later became a centralized location for government and industrial cooperation and development.

The purposes and plans for an interconnected system of utilities survived after the immediate military need waned. The question remained, however: who would facilitate the process, public or private entities?

Would government involvement in the utility industry through research and development continue even after the immediate wartime need faded?

Wilson Locks, Dam, and Powerhouse, constructed from 1918-1927, was the first hydroelectric and multipurpose project for the Corps of Engineers. Wilson Dam Project, Alabama. Date: Unknown. Photograph courtesy of the United States Army Corps of Engineers.

Women operators at powerhouse doing guard duty during World War I. Minidoka Project, Idaho. Date: 1918. Photographer: Unknown.

Grand Coulee Dam penstock installation. Columbia Basin Project, Washington. Date: May 7, 1940. Photographer: Unknown.

Colorado River, looking upstream near the Gila River Confluence, Arizona and Nevada. Date: September 17, 1992. Photographer: Carl Trulsen.

SUPERPOWER VERSUS GIANT POWER

In the 1920s, politicians and engineers referred to the process of integration of individual electrical distribution networks under private auspices as "Superpower." Superpower supporters advocated the creation of large, new powerplants tied to existing plants with transmission lines. The utility industry considered hydropower a vital component of the interconnected system since hydropower can be turned on and off immediately by regulating water flow to the turbines. It was, therefore, ideal for use in times of high electrical demand, known as "peaks."

Others, such as Pennsylvania Governor Gifford Pinchot and reformer Morris L. Cooke, felt publicly owned and managed municipal and cooperative utilities needed to be involved in order to include domestic and rural markets, which most private utilities considered unprofitable to service. The more publicly oriented power proponents referred to this latter version of system integration as "Giant Power." Giant Power aimed to pool regional power systems into a national network and promised to deliver electricity produced at Federal facilities more widely, including to rural areas, thus improving the quality of life for all people across the Nation. Critics denounced Giant Power as socialistic, while Superpower proponents continued to unite private utilities across the Nation.[62]

Concrete cones for draft tubes in powerhouse. Wilson Dam Project, Alabama. Date: Unknown. Photograph courtesy of the United States Army Corps of Engineers.

THE COLORADO RIVER

In the early 1920s, Reclamation entered the field of public power with its proposed hydroelectric development of the Colorado River Basin at the future site of Hoover Dam. However, before anyone could begin to use the resources of the Colorado River, the issue of ownership of the water in the river needed to be resolved.

The area through which the Colorado River flows is known as the Colorado River Basin. The basin is comprised of parts of seven states – Arizona, California, Colorado, New Mexico, Nevada, Utah, and Wyoming. The majority of runoff into the river comes from Colorado, New Mexico, Utah, and Wyoming. The logistics of dividing and utilizing a river that flows through such a large area challenged early Reclamation engineers, as well as the seven states within the river basin.

Because of its vast size, investigations of the Colorado River for irrigation development began as early as 1869, when John Wesley Powell traversed much of its length. Soon after Reclamation was created in 1902, Reclamation engineers began their own investigations of the river in an attempt to control flooding and to irrigate basin lands. Federal entities were not the only ones to set their sights on the waters of the Colorado. As early as 1905, William Mulholland, in the name of the city of Los Angeles, declared the river an ideal source of water for the impending growth of the area. The other six states also eyed the river for development, but none of them then possessed the resources to capitalize on the water. At the same time, the six states became increasingly more anxious about southern California's voracious appetite for water. In 1920, Arizona, Colorado, New Mexico, Nevada, Utah, and Wyoming banded together and invited California to join them in drawing up a compact to govern the use of the river in an attempt to put reasonable limits on California's appropriation of the Colorado River and allow development in the other six states.

In 1921, Congress passed a bill authorizing President Warren G. Harding to appoint a Colorado River Commission to negotiate a Colorado River Compact. At the same time, Arthur Powell Davis, then Reclamation's Director and Chief Engineer, issued a long-awaited report on potential development of the Colorado River. In his report, Davis advocated construction of a high dam in the Boulder Canyon area and the All-American Canal System leading to southern California. Davis's report proposed that those benefitting from the canal, primarily the growing agribusiness in the Imperial Valley, pay for its construction. The report also advocated financing of the dam through the sale of hydroelectric power.

President Herbert Hoover and party in Tunnel No. 2 during an inspection tour of Hoover Dam. Left to right: Construction Engineer W.R. Young, USBR; Mr. Ritchey, Secretary to the President; Chief Engineer R.F. Walter, USBR; Mrs. Hoover; President Hoover; Mrs. Wilbur; Secretary of The Interior R.L. Wilbur; E.O. Wattis, First Vice-President, Six Companies, Inc.; F.T. Crowe, General Superintendent, Six Companies, Inc. Boulder Canyon Project, Arizona, Nevada. Date: November 12, 1932. Photographer: Ben Glaha.

DEVELOPMENT OF THE HYDROELECTRIC POTENTIAL ON THE COLORADO RIVER

The high energy potential along the Colorado River provided another venue for conflict between the public and private power interests. Control of power development on the Colorado proved a major hurdle for those interested in harnessing it. Unfortunately for those involved, the public versus private power issue erupted on a national level just before negotiations for the Colorado River Compact (Compact) began.[63]

The scale and potential of the site chosen for Hoover Dam lent support to proponents of public power. Congress

proponents of public power. Congress realized that some projects were too large for private development, and Reclamation still needed power to repay costs of irrigation projects. In addition, the project would contribute to the booming city of Los Angeles, which made the project attractive to politicians. Many believed only the Federal government could undertake such a massive project. Conservationists endorsed the project as a multipurpose facility, while others believed such an influential public project would force down private rates under a "yardstick" principle.[64]

View of the All-American Canal. Boulder Canyon Project, California. Date: Unknown. Photographer: Unknown.

Davis believed that hydropower could finance the dam in one of two ways. Congress could appropriate the funds for construction of the dam and then "lease the power privileges" at the reservoir site to recoup most of its expenditure. If Congress opted not to appropriate funds for construction, Davis suggested that the "government should contract with any agency or agencies – public or private – that would build the dam themselves for the power to be developed."[65] Davis's recommendations directly addressed the national public versus private power debate: either the government would build the dam and develop the power, or a private entity would have the opportunity.

Though he presented both options for construction of the dam to Congress, Davis himself preferred that the Federal government build and own the dam. He felt that Congress would refuse to authorize such a move, largely due to the expense involved. Nevertheless, he remained confident that power sales would eventually repay construction costs of the entire project. Davis's recommendations to Congress met with mixed reaction. For the most part, its supporters and opponents fell naturally on either side of the ongoing public versus private debate.[66]

The Colorado River Compact

In 1922, while national interests continued to debate who should develop hydroelectric works around the country, the seven states in the Colorado River Basin returned to the issue of Colorado River water and began negotiating the Colorado River Compact.[67] Preferring to focus only on water, the states did not concern themselves with the development of hydropower during their negotiations. Instead, passage of the Boulder Canyon Project Act in 1928 relied on the resulting harmonious allocation of the river among the basin states.

The Colorado River Compact separated the Colorado River Basin into two smaller entities, the Upper (Colorado, New Mexico, Utah, and Wyoming) and Lower Basins (Arizona, California, and Nevada), with the dividing point located at Lee Ferry, Arizona. According to the Compact, each basin would receive 7.5 million acre-feet of water annually, with an additional one million acre-feet allotted to the Lower Basin from its tributaries. Representatives from the seven states signed the Compact in Santa Fe, New Mexico, on November 24, 1922. Before becoming binding and obligatory, the original terms of the Compact required that all seven states and Congress approve it. As a result, after its completion, the Colorado River Commission sent the Compact to the state legislatures and the Congress of the United States for approval.

Arizona refused to approve the Compact without a further division of the waters of the Lower Basin to keep California from taking the lion's share of the water allocation. The stalemate continued for six more years, when at the urging of the Upper Basin States, Congress ratified the Compact. The revised Compact limited California's allocation of the Lower Basin's share of the water to 4.4 million acre-feet per year. On December 21, 1928, Congress passed the Boulder Canyon Project Act, which included Congress's ratification of the Colorado River Compact and a provision for passage of the Compact after being approved by six of the seven states, provided one of the states was California. Six months later, on June 25, 1929, after six of the states ratified the Compact, President Herbert Hoover, in a public proclamation, declared the Boulder Canyon Project Act effective. Arizona chose not to ratify the Compact for nearly two additional decades.[68]

HOOVER DAM

With passage of the Boulder Canyon Project Act in 1928, Congress supported Davis's recommendation regarding construction of a Federal power facility along the Colorado River. In addition to making the Colorado River Compact binding, the Boulder Canyon Project Act authorized the Bureau of Reclamation to construct a high dam (Hoover Dam[69]) at Boulder Canyon or Black Canyon, and related irrigation works (the All-American Canal and Imperial Dam and Desilting Works), leading to the Imperial Valley in California. Reclamation engineers later moved downstream from Boulder Canyon to the alternate damsite, Black Canyon.[70] As Davis originally advocated, the authorization called for construction of a powerplant at the dam, with funds from the sale of surplus power being used to repay construction costs. Furthermore, the Boulder Canyon Project Act created a separate fund within the United States Treasury to finance project construction. The Colorado River Dam Fund is still active and is one of a few unique funds within the Treasury. All expenditures on the Boulder Canyon Project (BCP) – including those for construction, operation, maintenance, and repairs – come from this fund. At the same time, Reclamation returns all the revenues earned at Hoover Dam to this same fund. In this way, Congress was able to advance the funds for construction without overextending the Reclamation Fund.

Boulder City, Nevada, headquarters of the Boulder Canyon Project as seen from the air. Boulder Canyon Project, Arizona, Nevada. Date: December 13, 1936. Photographer: Unknown.

Reclamation leased the rights to generate power at Hoover to outside utility companies – including Southern California Edison, Los Angeles Bureau of Power and Light, The Metropolitan Water District of Southern California (MWD), Southern Sierras Power Company, and the states of Arizona and Nevada – to distribute to the surrounding areas. Reclamation constructed the generators and switchyards while the utilities built the related distribution facilities, which included transmission lines and substations. Largely because of its hydropower component, the BCP was Reclamation's and the Nation's first great multipurpose project.[71]

An aerial view of Hoover Dam, with Lake Mead and Fortification Mountain in the background. Boulder Canyon Project, Arizona, Nevada.
Date: March 1996. Photographer: Andrew Pernick.

On September 11, 1936, as part of ceremonies for a World Power Conference, President Franklin Delano Roosevelt (FDR) pushed a golden telegraph key in Washington, D.C., starting the first generator at Hoover Dam. The first commercial delivery of power occurred the following month when regular service to Los Angeles began.[72] By 1939, Hoover had nine of its seventeen commercial generators online and held the distinction of being the world's largest hydroelectric facility.[73]

Hoover Dam changed the way Congress and the public viewed Reclamation. Until authorization of the BCP, Reclamation relied almost exclusively on public land sales and various other legally mandated revenues for funding. Hoover Dam's reliance on power revenues as a way to repay construction and operating costs specifically acknowledged Reclamation's growing role in the development of hydropower, as well as the growing importance of public works projects.

The BCP Act heralded a new direction in Federal construction of power facilities in the West. Prior to the authorization and construction of the powerplant at Hoover Dam, Reclamation had constructed hydropower plants with a total generating capacity of more than 37,000 kW. Prior to 1940, the generators at Hoover alone had a capacity of more than 746,000 kW. Total generation capacity was 1,340 megawatts after the last generator went online in 1961. Total generating capacity increased to 2,078 megawatts during the 1982 to 1993 rewinds. Private interests owned and controlled plants with a combined generating capacity of just over two and a half million kilowatts. With Hoover, the Federal government entered the power business in a significant manner and changed the relationship between public and private power.

No longer would private interests control the power market in the West; the Federal government was now poised to become a significant presence. At the same time, Federal licensing of powerplants on the Colorado became dependent on Hoover. There had to be enough water to power the massive generators at the dam, and the Federal Power Commission took that into account when private entities requested permission to construct new facilities on the river. Between the capacity of Hoover's generators and the role of the FPC in development of hydropower sites, Federal control of the Colorado River was ensured.[74]

Hoover, the largest project undertaken by Reclamation to that date, formed a large reservoir (Lake Mead) in which the three Lower Basin States could store water. In addition, the Boulder Canyon Project Act heralded a new era in dam building. Advancements in dam technology, design, and construction developed on earlier Reclamation projects made construction of Hoover feasible. At the same time, construction of Hoover Dam paved the way for additional projects on the same scale. Previously, Reclamation focused its efforts on small-scale projects that provided water to surrounding irrigation lands. On the other hand, the Boulder Canyon Project, the first truly multipurpose project, did not emphasize irrigation development; rather, it provided flood control, the storage and controlled release of water for municipal and industrial use, as well as hydropower production. Irrigation lands fell under the complementary All-American Canal System portion of the project.

Not surprisingly, in many ways, California proved to be the ultimate beneficiary of the Boulder Canyon Project Act. At the start of project construction activities in 1931, California was the only Lower Basin State in a position to capitalize on the future water supply. None of the other states had the necessary water works developed or a need to take advantage of the waters to be stored in Lake Mead. In addition, the All-American Canal led straight to southern California, whose growing industrial activity provided a ready-made market for the power generated at Hoover.[75] The growing populations of Phoenix, Arizona, and Las Vegas, Nevada, also offered suitable

Bureau of Reclamation office buildings nearly completed, in Boulder City, Nevada. Boulder Canyon Project, Arizona, Nevada. Date: c. 1935. Photographer: Unknown.

Drillers – high-scalers – at work on the canyon wall above the powerplant location at Hoover Dam. Boulder Canyon Project, Arizona, Nevada. Date: April 10, 1933. Photographer: Ben Glaha

locations for use of project water and power, and both cities capitalized on the opportunity for growth.

In the midst of preparing to build the project, the United States entered into the Great Depression. Reclamation projects provided much-needed jobs for unemployed Americans. Responding to the economic crisis, President Franklin D. Roosevelt capitalized on successful projects like Hoover to promote additional public works projects. Construction of Hoover Dam became one of the best known Depression Era accomplishments in the United States and captured America's imagination. Construction of Hoover began only after the Depression had alarmed Americans, and the fact that authorization of the project occurred before the start of the Depression did not matter when it came to its use as positive promotion of public works.[76]

View of downtown Las Vegas, Nevada, looking east from corner of Fremont and Main Streets. Date: July 17, 1945. Photographer: William S. Russell.

Footnotes for Chapter 4: The Rise of Public Power

[53] Nye, *American Technological Sublime*, 135, 140-143.

[54] Nye, *Consuming Power*, 249-51; See also Hughes, *Networks of Power*.

[55] In 1883, Massachusetts legislation placed the natural gas business under supervision. Suellen M. Hoy, Michael C. Robinson, and Ellis L. Armstrong, *History of Public Works in the United States* (Chicago: American Public Works Association, 1976), 343.

[56] Jay L. Brigham, *Politics and Power: The Fight For Electricity in the West, 1902-1932* (Lawrence, Kansas: University of Kansas Press, 1998), 9-15.

[57] Hoy, Robinson, and Armstrong, 347; Brigham, 16-17.

[58] Regional systems developed from 1935 to 1960.

[59] An 1899 Supreme Court decision decreed that Congress had the constitutional right to regulate commerce between states and nations through waterways. Congress revised the Federal Power Act in 1930, creating the Federal Power Commission as an independent, full-time body. In the Department of Energy Organization Act of 1977, Congress renamed the Federal Power Commission as the Federal Energy Regulatory Commission (FERC).

[60] Brigham, 21.

[61] See Thomas P. Hughes, *American Genesis: A Century of Invention and Technological Enthusiasm, 1870-1970* (New York: Viking University Press, 1989); Richard White, *The Organic Machine* (New York: Hill and Wang, 1995), 50-51.

[62] Richard Rudolph and Scott Ridley, *Power Struggle: The Hundred-Year War Over Electricity* (New York: Harper and Row, 1986), xi, 28, 51.

[63] Toni Rae Linenberger, "A Dam for All Seasons: Hollywood, the Bureau of Reclamation, and Construction of Parker Dam" (M.S. Report, Utah State University, 1998), 7-8.

[64] Brigham, 46-47, 50, 62; John Richard Waltrip, *Public Power During the Truman Administration* (New York: Arno Press, 1979), 6-14.

[65] Hundley, *Water and the West*, 113-114.

[66] Hundley, *Water and the West*, 114.

[67] Norris Hundley, jr., *The Great Thirst: Californians and Water, 1770s-1990s* (Berkeley: University of California Press, 1992), 211-213.

[68] Despite the inclusion of a clause further appropriating the waters of the Lower Basin between Arizona, California, and Nevada, Arizona still felt that the Compact favored California. Arizona further argued that if they did not ratify the Compact, then it would not be binding to the state. In 1944, after passage of the Mexican Treaty of 1944, requiring an annual delivery of 1.5 million acre-feet of water from the Colorado River to Mexico, Arizona ratified the Colorado River Compact. At the same time, Arizona contracted for delivery of the 2.8 million acre-feet of water

allotted to the state in the Boulder Canyon Project Act. Reclamation's formulation of several preliminary plans for the Central Arizona Project (CAP) probably played a role in Arizona's sudden willingness to accept the Compact as written.

For more information on the Colorado River Compact, see Norris Hundley, jr. "The Politics of Reclamation: California, the Federal Government, and the Origins of the Boulder Canyon Act – A Second Look," *California Historical Quarterly* 52 (Winter 1973): 292-325; Norris Hundley, jr, "Clio Nods: *Arizona v. California* and the Boulder Canyon Act – A Reassessment," *Western Historical Quarterly* 3 (January 1972): 17-51.

[69] In 1930, Secretary of the Interior Ray Lyman Wilbur officially named the dam Hoover Dam in honor of former President and chief negotiator of the Colorado River Compact of 1922, Herbert Hoover. In 1933, Secretary of the Interior Harold Ickes designated the structure Boulder Dam; however, he never officially renamed the structure. The Nation at large held Hoover responsible for the Great Depression, and the name change reflected the national mood, as well as Ickes' own dislike of the former President. On April 30, 1947, Congress permanently restored the name Hoover Dam. In this study, for purposes of consistency, the structure will be referred to by its current name.

[70] Linenberger, *A Dam for All Seasons*, 8-9.

[71] Hundley, *The Great Thirst,* 220; United States Department of the Interior, Bureau of Reclamation, *Hoover Dam* (Washington, D.C.: Government Printing Office, 1985), 40.

[72] Robinson, 54.

[73] Joseph E. Stevens, *Hoover Dam: An American Adventure* (Norman: University of Oklahoma Press, 1988), 259; *Hoover Dam*, 6-7.

[74] David P. Billington, Donald C. Jackson, and Martin Melosi, *Federal Multipurpose Dams: The U.S. Army Corps of Engineers and the Bureau of Reclamation in the Era of Big Dams* (Draft Manuscript: December 22, 2000), 4-59, 5-37.

[75] Linenberger, *A Dam for All Seasons*, 9-10.

[76] Linenberger, *A Dam for All Seasons*, 12-13.

On this day, a 505 foot by 225 foot American flag was flown from the downstream face of Hoover Dam. The flag itself weighed 3,000 pounds and was hoisted by three separate cables located at the top of the dam. The flying of the flag was part of a ceremony in which the 1996 Summer Olympic torch was carried across Hoover Dam. Boulder Canyon Project, Arizona, Nevada. Date: May 1, 1996. Photographer: Andrew Pernick.

Seeking employment at Hoover Dam during the Depression. Boulder Canyon Project, Arizona, Nevada. Date: Unknown. Photographer: Unknown.

Entrance to Boulder City, Nevada. Boulder Canyon Project, Arizona, Nevada. Date: 1932. Photographer: Teddy Fenton.

CHAPTER FIVE

In the 1920s, Americans clung to traditional ways and values but still looked optimistically toward new ideas and technology to solve problems. At the decade's end, the prosperity enjoyed after World War I came to an abrupt end. Since people looked to the government to modernize in the 1920s, it is not surprising that advocates for public power gained the upper hand when the stock market crashed in 1929, sending the entire Nation spiraling into severe economic distress and into a recession. Over the next two months, national production declined at an annual rate of twenty percent, wholesale prices declined seven and one-half percent, and personal income declined five percent. On Tuesday, October 24, 1929, the New York stock market crashed, signaling a prolonged worldwide economic depression.

PUBLIC ATTITUDES TOWARDS ELECTRICITY

Many people believed that electricity should be available to everyone, regardless of their class or economic status. Electricity could be used to make life easier and better for everyone. This belief, coupled with the necessity of using electricity immediately after it was generated, prompted the launch of a major campaign to publicize and sell electricity. The electrification program was a giant advertisement for the good that technology could do if people had electricity and used it. It was a clean, economical, cheap, flexible, and accessible energy source that appealed to almost everyone.[77] Domestic modernization was a part of the reform tradition begun in the 1920s and further articulated in the New Deal. However, complete electrical modernization across class lines would not really occur until President Truman's "Fair Deal" programs focused on equal opportunity housing, including a standard of electrical modernization, after World War II.[78]

Part of the positive public interest in electricity stemmed from intellectual movements. One such movement, the technocracy movement, argued that technology would solve problems, rescue Americans from the Great Depression, and return the Nation to its former prosperity. Technocrats also agreed with the conservationists, who supported the efficient use of natural resources and their conversion into usable forms of energy.[79]

Those who feared the mechanization of humanity countered the arguments of the technocrats. This public ambivalence over technology manifested itself in government policy that professed to minimize or control the negative effects of technology, but used technological change to aid and enhance social processes and problems. Presidents Calvin Coolidge, Herbert Hoover, and Franklin Roosevelt promoted technology as an agent of positive and needed change.[80]

ELECTION OF FRANKLIN DELANO ROOSEVELT AND THE NEW DEAL

By the time of the 1932 Presidential elections, voters looked toward more government involvement than Herbert

"Builders at Grand Coulee No. 9." Construction workers at the Third Powerplant, Grand Coulee Dam. Columbia Basin Project: Third Powerplant, Washington. Date: Unknown. Artist: Anton Refregier.

President Franklin D. Roosevelt at the dedication of Hoover Dam. Boulder Canyon Project, Arizona, Nevada. Date: September 30, 1935. Photographer: Unknown.

Grand Coulee construction engineer Frank Banks and President Franklin D. Roosevelt at Grand Coulee Dam. Columbia Basin Project, Washington. Date: October 2, 1937. Photographer: Unknown.

Hoover was willing to provide. Many prominent commentators of the day, such as philosopher John Dewey, considered public power a critical factor in the election that brought victory to Franklin Delano Roosevelt. The concept that electricity was essential to modern life emerged as part of the public psyche. The recognition that many Americans did not have access to electricity made this a major political issue of the day.[81]

For years, many Senators had been fighting for public power and against the continuing consolidation and influence of the private electrical utility industry. Most of these Senators' constituents needed aid on a scale that necessitated dramatic action and significant funding that only the Federal government could or would provide.

FDR's innovative ideas about natural resources and natural resource development influenced his cabinet choices and made him the perfect President to further the public power movement. President Roosevelt's Secretary of the Interior Harold Ickes had long been committed to public power and had fought private monopolies. Felix Frankfurter, Roosevelt's chief advisor on utilities, declared "hydroelectric power raises without doubt the most far-reaching social and economic issues before the American people, certainly for the next decade."[82]

According to Frankfurter, the issue of hydropower eclipsed even the Depression in terms of importance. Yet, many historians believe that government immersion in public power issues and hydropower development would have receded into the background without the Depression.

With the failure of local employers or institutions to provide aid in difficult times, Americans looked to the Federal government, which provided many people with their only socio-economic

Standing on a hill near Tule Lake, California, overlooking project lands, Robert Elmer Metz, whose homestead application was drawn, is congratulated by Frederick Lehman, while Mrs. Metz looks on. Klamath Project, California, Oregon. Date: December 18, 1946. Photographer: Ben D. Glaha.

assistance and support. President Roosevelt fertilized this expectation with his philosophy of aggressive action. During his four terms in office, and particularly through his 1930s New Deal programs, he expanded the power, duties, and expectations of the Executive Branch of government. Individualism and private worlds gave way to collective social action and the public arena. Faced with unemployment and poverty, FDR's paternalistic leadership was just what people needed and desired during a time of crisis. Americans sought creative and cooperative solutions. The New Deal years introduced the idea of the government caring for its citizens.[83]

NEW DEAL PUBLIC WORKS PROGRAMS

Although ideas to federally subsidize research for a technological remedy to the Great Depression did not provide a solution to unemployment, the New Deal construction projects were especially effective instances where technological and employment needs complemented one another. After the Boulder Canyon Project Act in 1928, Congress approved the initial units of the Central Valley Project in California in 1935, the Columbia Basin Project (Grand Coulee Dam) in Washington, also in 1935, and the Colorado-Big Thompson Project in Colorado two years later. Each of these projects symbolized a faith that engineering could harness nature and benefit humanity and society.

The first acts of the New Deal in 1933 included the Tennessee Valley Authority and the National Industrial Recovery Act of 1933 (NIRA) as part of Roosevelt's famous "first 100 days" of legislation. The TVA proposed to further develop the Muscle Shoals facilities and the Tennessee Valley, an economically promising but poverty-stricken region, through hydroelectric power, flood control, and soil conservation. The NIRA promoted recovery by condoning certain monopolistic practices among major industries.[84] In June of 1933, NIRA also authorized the Public Works Administration (PWA) to assist the financing of public projects. It provided employment relief and carried the authority to fund the construction of municipal electric systems and public power districts. The same act allowed the government to initiate public works, including Bonneville Dam in Oregon, Fort Peck Dam in Montana (both Army Corps of Engineers' projects), and California's Central Valley Project. The government immediately employed the workers, who began construction on these projects.

The promise of enormous hydroelectric potential and development for the Pacific Northwest was similar to that in the Tennessee Valley, and it launched another local battle over control and operation. Washington state newspaper *Wenatchee World* editor Rufus Woods supported big government dams on the Columbia River if they promoted growth for the Northwest. He and others considered the Columbia River one of the greatest sources of hydroelectric power in the country. Willis Batcheller, a Seattle engineer, first envisioned Grand Coulee Dam as a facility to generate hydropower, while engineer and lawyer James O'Sullivan felt the dam could go far in financing irrigation, still the primary purpose of all Reclamation dams. In melodramatic tones, Woods' newspaper stories in the early 1920s drummed up public support by casting opponents of a dam at Grand Coulee as villains. But no one wanted or needed the electricity, particularly locally, and boosters had to wait for the Great

Aerial view of the downstream face of Grand Coulee Dam. Columbia Basin Project, Washington. Date: June 27, 1978. Photographer: H.S. Holmes.

Shasta Dam under construction. Central Valley Project: Shasta/Trinity River Division, California. Date: Unknown. Photographer: Unknown.

Governor C.D. Martin tripping the concrete bucket at the first official pour ceremony at Grand Coulee Dam. Columbia Basin Project, Washington. Date: November 1935. Photographer: Unknown.

Pumping plant at Grand Coulee Dam. Columbia Basin Project, Washington. Date: Unknown. Photographer: E. Hertzog.

Depression when electricity could be recast as a means of transforming society and putting people to work.[85] After passage of the TVA, Major John Butler of the Corps began a survey of the Columbia River Basin, which he conducted from 1928 to 1931. His report favored building a large dam at Grand Coulee. Like many others of its kind, including Hoover, the proposed dam would provide its greatest immediate benefits by providing jobs in a time of massive unemployment.

The PWA became the Bureau of Reclamation's primary source of funding from 1933 to 1943. Generous allocations, including sixty-three million dollars for construction of Grand Coulee Dam on the Columbia Basin Project, allowed Reclamation to arguably become the most influential Federal bureau in the West.[86] In 1939, Congress passed one of Reclamation's most important pieces of legislation that enhanced its power capability in the name of conservation. The Reclamation Project Act of 1939 stipulated that power benefits be considered when determining the economic feasibility of multipurpose projects involving irrigation. In fact, sections of the 1939 Act supported the use of power revenues, when available, to support repayment of construction costs on Reclamation projects.[87] The entry of the government into building multipurpose projects established Reclamation as a primary influence in the modernization and industrialization of the American West.

THE RISE OF PUBLIC POWER

The Great Depression helped destroy utility holding companies, whom many blamed for the economic crisis. In the golden age of speculative stock trading, utility companies led the stock

Football game at Coulee Dam CCC Camp. Columbia Basin Project, Washington. Date: October 1935. Photographer: K.S. Brown.

J.P. Henry, of the Nampa, Idaho, plant of the Amalgamated Sugar Company, and Ellen Dalton watch as M. Oberat brings in another 700 pounds of beet sugar tagged for Mrs. Housewife, USA. Boise Project, Idaho. Date: February 4, 1946. Photographer: F.B. Pomeroy.

market. Samuel Insull, Thomas Edison's former administrative secretary and manager, emerged as the most popular businessman in the 1920s, but later public opinion condemned him as a notorious villain of self-serving and economically destructive practices. As a preventive measure against the future development of holding companies, Congress passed the Public Utility Holding Company Act of 1935 (PUHCA). The act allowed private industry to control the majority of the electrical industry but encouraged system integration and solidified Federal involvement in that aspect of the American economy through strict regulation and geographic restriction.[88]

The following year, 1936, the New Deal dealt a further blow to private industry with passage of the Rural Electrification Act, which made power available to rural Americans across the country who had been unable to obtain electricity from private utility companies. The Rural Electrification Administration (REA) offered Federal loans to states, public utilities, electrical cooperatives, and municipalities to build power lines to farms and ranches to make them more efficient, productive, and raise the quality of life for their residents. The local agencies purchased power generated by a municipal plant, Reclamation, or a local power company. Reclamation hydropower already benefitted rural residents in a number of ways. It produced revenue directed toward repayment of irrigation facilities, and it made irrigation of elevated lands easier through pumping. With REA's help, more rural people could enjoy the conveniences brought by electricity to their farms and homes. Historian William E. Leuchtenburg even suggested that "perhaps no single act of the Roosevelt years changed more directly the way people lived than the President's creation of the Rural Electrification Administration."[89]

The Flood Control Act of 1936 authorized the Army's Chief of Engineers and the Secretary of War to decide where to build facilities for electrical power development at dams primarily designed for flood control purposes. In 1937, the Bonneville Project Act furthered the influence of public power and sought to prevent private interests from forming regional monopolies through electrical transmission systems. Upon the recommendation of the Pacific Northwest Regional Planning Commission, Congress created the Bonneville Power Administration (BPA), an independent administration under the Department of the Interior. BPA controlled the marketing, sale, and distribution of power from Bonneville and Grand Coulee Dams, which continued to be operated by the Corps and Reclamation, respectively.[90] The policies and structure that public power-related New Deal programs put in place had a lasting influence on American government and the continued growth of Reclamation's hydropower program during and after World War II.

Third Powerplant at Grand Coulee. Columbia Basin Project: Third Powerplant, Washington. Date: Unknown. Photographer: Unknown.

Night view of Shasta Dam operations, as seen from point at left end of excavation near crest elevation. Central Valley Project, California. Date: March 11, 1941. Photographer: Ben D. Glaha.

This family of migrant farm laborers from Ardmore, Oklahoma, camped along Highway 99 just south of Tulare, California, as they headed north to Modesto to "work in fruit." Date: July 7, 1944. Photographer: Ben D. Glaha.

Footnotes for Chapter 5: Reclamation and the Great Depression

[77] Nye, *Consuming Power*, 171-173.

[78] Ronald Tobey, *Technology as Freedom: The New Deal and the Electrical Modernization of the American Home* (Berkeley: University of California Press, 1995), 1-6; Brigham, 20; Gene Tollefson, *BPA and the Struggle for Power at Cost* (Portland, Oregon: Bonneville Power Administration, 1987), 75.

[79] "Technocracy, Inc.," Pamphlet (Continental Headquarters), March 1933, <www.technocracy.org> accessed July 15, 1998.

[80] Carroll Pursell, "Government and Technology in the Great Depression," *Technology and Culture* 20:1, 162-174; Nye, *American Technological Sublime*, 6.

[81] Tobey, 1-6; Brigham, 20; Tollefson, 75.

[82] Tobey, 55.

[83] For more on this era, see Lizabeth Cohen, *Making a New Deal: Industrial Workers in Chicago, 1919-1939* (Chicago: Cambridge University Press, 1990); William E. Leuchtenberg, *Franklin D. Roosevelt and the New Deal: 1932-1940* (New York: Harper and Row Publishers, 1963); Roderick Nash, *The Nervous Generation, 1917-1930* (Chicago: Elephant Paperbacks, Ivan R. Dee Publishers, 1970); Robert S. McElvaine, *The Great Depression: America, 1929-1941* (New York: Times Books, 1993, second edition); Rudolph and Ridley, 55, 62.

[84] In 1935, the Supreme Court declared NIRA unconstitutional.

[85] White, 52-58.

[86] Robert E. Ficken, *Rufus Woods: The Columbia River and the Building of Modern Washington* (Pullman, Washington: Washington State University Press, 1995), 135-137; Richard Lowitt, *The New Deal and the West* (Bloomington: Indiana University Press, 1984), 223-228.

[87] United States Department of the Interior, *Annual Report of the Secretary of the Interior* (Washington, D.C.: Government Printing Office, 1939), 195.

[88] Rudolph and Ridley, 53; <http://www.townhall.com/pff/pucha/part2.html> accessed August 4, 1998.

[89] Leuchtenburg, 157-158.

[90] The first bill that proposed the distribution of power at Bonneville Dam, introduced in 1935, would have created the Columbia Valley Authority, modeled after the TVA. Also that year, legislation suggested that the Corps and Reclamation individually market their power. The last idea, to link the Bonneville and Columbia Basin systems regionally by appointing a separate government bureau to manage it, became the Bonneville Power Administration. For more information on the formation and function of the BPA, see discussion on regional systems. Vera Springer, *Power and the Pacific Northwest: A History of the Bonneville Power Administration* (United States Department of the Interior, 1976), 29-41; White, 64-65.

**Sergeant H.E. Whitlatch, U.S. Army, on guard duty at diversion tunnel. Central Valley Project: Kennett Division, California.
Date: December 11, 1941.
Photographer: T.B. Gibson.**

CHAPTER SIX

On the night of December 7, 1941, following the Japanese bombing of the naval base at Pearl Harbor, Hawaii, Secretary of the Interior Harold L. Ickes issued an Interior-wide order instructing all employees to mobilize the country's natural resources for the purpose of war. War efforts in the United States, however, began long before Japan's strike on the United States forces at Pearl Harbor. After Germany and the Soviet Union signed the Hitler-Stalin pact in August of 1939, both Europe and the United States started gearing up for what promised to be a full-scale conflict. On September 1, 1939, Hitler invaded Poland, beginning World War II in Europe. American war production efforts, including aircraft construction and anti-Fascist propaganda, began as early as 1939.

Members of Company 4821 on parade at Powell, Wyoming. Shoshone Project, Wyoming. Date: May 10, 1941. Photographer: J.E. Mowry.

PRE-WAR PREPARATION

On the eve of the second World War, New Deal legislation came closer than ever to constructing a national power policy. The high demand for power during the first World War had taken everyone by surprise, and the country did not want to be caught unprepared as the possibility of a second World War loomed. As a result, in 1938, President Roosevelt instructed the Federal Power Commission and the War Department to cooperatively survey the Nation's wartime power capacity.[91]

In a joint memorandum, the FPC and the War Department concluded that the government's public power agencies would have to collaborate with private utilities in a wartime emergency. Each entity then individually formulated a program. Encouraged by the Public Utility Holding Company Act of 1935, FPC engineers proposed development of an interconnected network comprised of high-capacity transmission lines and existing regional systems, like Niagara Falls and the Tennessee Valley, to expand hydro and steam power facilities. War Department engineers, on the other hand, recommended eliminating such a costly high-voltage network and increasing the Nation's generating capacity by constructing additional steam plants. The FPC's interconnection plan prevailed. On September 1, 1938, Roosevelt

Aerial view of BMI plant in Henderson, Nevada.
Date: November 16, 2001.
Photographer: Andrew Pernick.

established the National Defense Power Committee to create and put more formal plans in place. Many legislators continued to worry about the ramifications of a wartime power development program during peacetime. How would excess power and an interconnected transmission system affect national public policy or the national power policies created by New Deal legislation? Who would control the generation of power during peacetime? What would happen to all of the country's preparations if the Nation never went to war? Men like Secretary of the Interior Ickes worried about sacrificing the public power program developed during the New Deal. Regardless, war preparation efforts required the support of private industry, which again underestimated wartime power needs.[92]

America's participation in World War II applied enormous pressure to Reclamation's hydroelectric power capability. In 1940, the Axis Nations had fifty percent more available power than the United States. Secretary Ickes addressed the demand for additional power in a 1942 statement on "The War Program of the Department of the Interior." As a result, Interior's wartime budget called for an additional fifty-six billion dollars to generate 154 billion kilowatt hours to manufacture airplanes, tanks, guns, warships, and fighting material to equip the military. Processing aluminum to produce 60,000 airplanes alone required an additional 8.5 billion kilowatt hours. These requirements exceeded the total combined production of all electric utilities in 1940. To meet the increased demand for power, Interior planned to triple its 1941 power production. Secretary Ickes called on all of the power-producing bureaus within Interior to increase the output from powerplants in the western states to 35.5 billion kilowatt hours in five years and to efficiently allocate energy among war industries. Secretary Ickes also specifically recommended that Reclamation begin construction of additional hydro and steam plants to add 1.48 million kilowatts, with an output of nine to ten billion kilowatt hours, by 1945. The war program and, by extension, Reclamation would need to "pull out all the stops" in the conservation and diversion of water, and in the generation and expansion of energy.[93]

Reclamation was often the major producer of electricity in areas where major underdeveloped natural resources were located. The planning of multipurpose irrigation facilities before the war allowed Reclamation to enlarge the power facilities of already approved projects. As a result of such projects, the Federal government's power production increased twenty-five percent by 1940. Recently completed powerplants at Hoover Dam, Shasta Dam on the Central Valley Project in California, and Grand Coulee Dam together could produce more than seven billion kilowatt hours annually. The Hoover Dam Powerplant supplied the energy for aircraft plants in the Los Angeles area and the largest magnesium plant in the world near Henderson, Nevada.[94]

BASIC MAGNESIUM INCORPORATED

The first generator at Hoover, unit two on the Nevada side, began generating commercial power on October 26, 1936. Two additional units went online later the same year. By the end of 1939, Hoover had nine generators in operation, making it the largest hydroelectric facility in the world, a distinction it held until Grand Coulee surpassed it in 1949.[95] The expanding power production at Hoover conveniently coincided with the booming war and probably attracted wartime industries to Nevada and southern California.

The Basic Magnesium, Incorporated, (BMI) plant adjacent to Henderson, Nevada, was a result and direct beneficiary of construction of hydropower facilities at Hoover. The promise of abundant water and power from Hoover Dam lured them to build the plant in its Henderson location. BMI was the first such plant; a sprawling light-metals complex soon sprang up next to the new town of Henderson. BMI became one of the most significant wartime industries to receive Hoover-generated energy. Built by the Defense Plant Corporation, with funding from the War Assets Administration, the government plant played a significant role in the defense industry. Producing magnesium for aircraft frames and incendiary bombs, the plant used one-fourth of the energy generated at Hoover in 1943. Soon after construction of BMI, the United States Army Air Corps (later the United States Air Force) established a large training school and gunnery range in the area (the Las Vegas Army Air Corps Gunnery School, now Nellis Air Force Base, home of the Thunderbirds),[96] again taking advantage of the nearby source of water and power and bringing even more defense dollars into the Las Vegas area.[97]

THE WAR EFFORT IN SOUTHERN CALIFORNIA

One of the first bedroom communities outside of Los Angeles, Burbank's origins date back to 1887. In 1911, the same year as its incorporation, the city connected to the Pacific Electric streetcar system, which brought additional urban settlement and industrial activity. At the same time, a new industrial infrastructure based on oil, movies, and the budding aircraft industry emerged throughout various settlements in the Los Angeles region. In 1926 and 1927, Lockheed, an aircraft manufacturer, and Warner Brothers, a Hollywood film studio, established themselves in Burbank; they later became two of Burbank's most prominent companies.

Lockheed, founded in 1916 and incorporated a decade later, built its major facility in Burbank in 1927. A year later, developers selected a Burbank site to house a motion picture studio, which Warner Brothers soon acquired.[98] Both companies' presence in the small town forced Burbank to focus on its infrastructure and its ability to meet anticipated growth, with

The USS John C. Stennis (CVN 74), assisted by tug boats as it maneuvered in San Diego Bay. Date: Unknown. Photograph courtesy of the United States Navy.

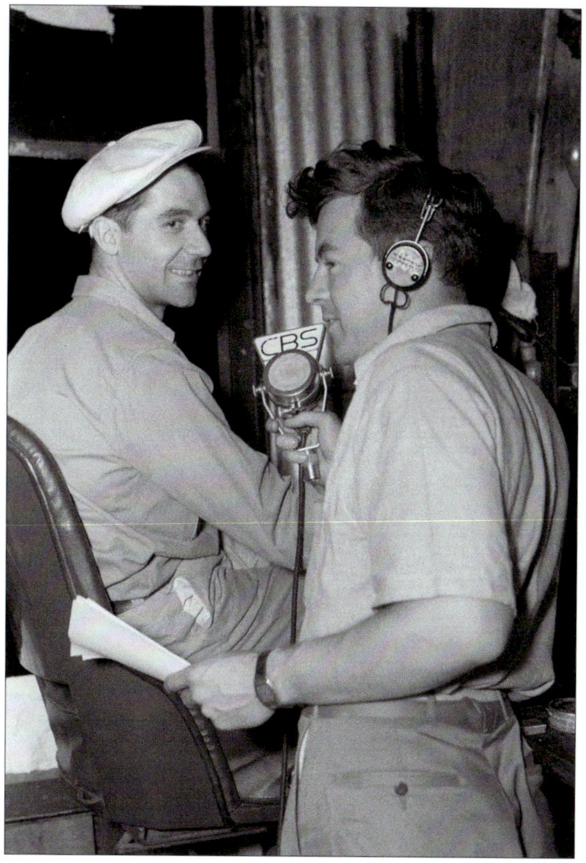

"Happy" Hepner, hoist operator for Pacific Constructors, Inc., being interviewed by Tro Harper, CBS announcer, from Shasta Dam. Date: August 17, 1941. Photographer: T.B. Gibson.

View of the Nevada generators inside the Hoover Dam Powerhouse. Boulder Canyon Project, Arizona, Nevada. Date: January 1996. Photographer: Andrew Pernick.

Aerial view of Parker Dam and Lake Havasu. Parker-Davis Project, Arizona, California. Date: 1997. Photographer: Andrew Pernick.

water and energy as a central concern. Because they did not annex themselves to the city of Los Angeles, Burbank was ineligible to receive water from the Los Angeles Aqueduct. To address the growing need for water and energy, Burbank, and several other cities within the region, joined the quest to receive additional resources from the Colorado River by forming The Metropolitan Water District of Southern California.

MWD, formed in 1928, was organized for the purpose of supplying water to the cities within its boundaries. To achieve its goal, the first objective was construction of the Colorado River Aqueduct. Leading from the Colorado River near Parker, Arizona – now the location of Lake Havasu behind Parker Dam – the 242 mile long aqueduct delivers water to the cities of southern California. To assure the delivery of water, MWD contracted with Reclamation for the construction of Parker Dam, downstream of Hoover.

Though water proved a nice boon, the city of Burbank primarily focused on the electricity generated at Hoover Dam to supply Lockheed's and Warner's growing energy needs. In expectation of signing contracts with the Federal government for energy generated at Hoover, the city bought the Burbank facilities of Southern California Edison in 1934. Three years later, the city contracted with the government for Hoover energy. As such, Burbank holds the distinction of being one of the first recipients of energy generated at Hoover Dam.

Energy supplied by Hoover allowed both companies to expand, and both, in turn, became huge World War II industries. After declaring bankruptcy and reorganizing as part of the consolidation of the aircraft industry during the Depression, Lockheed re-emerged as a major energy consumer during the War. Despite a brief postwar cutback, Lockheed continued to grow throughout the Korean conflict and Cold War.[99] Lockheed was not the only aircraft manufacturer to establish plants in southern California to take advantage of the growing power development on the Colorado River. Douglas, Vultee, North American, and Northrup, all with plants in southern California, also played a significant role in United States war production efforts.[100] Three of the southern California aircraft plants – Douglas, Lockheed, and North American – built approximately 62,000 fighters, bombers, and cargo planes, a fifth of the Nation's entire aircraft production between 1941 and 1945.[101]

Warner Brothers Studio naturally played a far different role during World War II. As early as 1939, Hollywood, the self-proclaimed home of the film industry, began producing anti-German propaganda films. Warner Brothers was the first of the major studios to take a stand against Hitler on film; the other studios refrained from overt propaganda due to economic concerns. The studios cited the Hollywood Production Code[102] and concern for the Jewish people in Europe as official reasons for their conciliatory attitude.

Warner Brothers released the first anti-Fascist film, *Confessions of a Nazi Spy*, in 1939. Warner Brothers' emphasis on anti-Fascist propaganda eventually convinced the other major studios to join the fray.[103] Many of the other studios also benefitted from power produced at Hoover and later at Parker Dam and Powerplant 167 miles downstream from Hoover.

In 1941, the bombing of Pearl Harbor erased all restraint, and widespread film propaganda began in earnest. The influence of the film industry had already become so significant by the end of 1941 that the government quickly declared movies an "essential industry" much like coal and steel production, shipbuilding, and food processing. In his State of the Union address to Congress in January of 1942, President Roosevelt explained that six areas existed around which film propaganda should be based: the issues of war (why we were fighting), the nature of the enemy, our allies, the production front, the home front, and the fighting forces. Hollywood took its Presidential directive seriously and began producing some of America's best known films including *Casablanca*, *Suspicion*, and *To Be or Not to Be*.[104]

After the bombing of Pearl Harbor, Hollywood and its studios found themselves in a rather unique position. Southern California housed two-thirds of the Nation's major aircraft plants, which suddenly found themselves wary of possible Japanese attack. To counteract this new fear, an Army guard unit moved into the Walt Disney Studio to protect the Lockheed aircraft plant, located next door. The nearby Douglas aircraft plant took a slightly different approach. They hired Warner Brothers' special effects technicians to camouflage their buildings to make them unrecognizable from the air. The camouflage proved so successful, Lockheed asked the Warner Brothers' technicians to do the same for their plant. Hoover's designation as a prime strategic target suggested a similar solution.[105]

Because of Hoover's proximity to the California coast and its production of power and storage of water, Reclamation feared that it would be targeted for enemy attack. To combat the threat of aircraft attack, workers at Hoover designed a camouflage system of their own. The Hoover design consisted of two separate tarps – a large blue one to cover the dam itself, allowing it to blend into the river, and a second tarp designed to look like the dam placed farther upstream. Even if one of the Axis Countries launched a successful attack, they might not hit the structure. The Army also moved into Boulder City to man the checkpoints and gun emplacements that guarded the dam from attacks by Japanese aircraft or saboteurs.[106] Hoover was not the only dam placed under guard during the war. Other structures along the Colorado River also went under guard, including Parker Dam just downstream on the Arizona-California border. However, the Army's stay at Parker was short because the unit stationed there soon shipped out overseas.

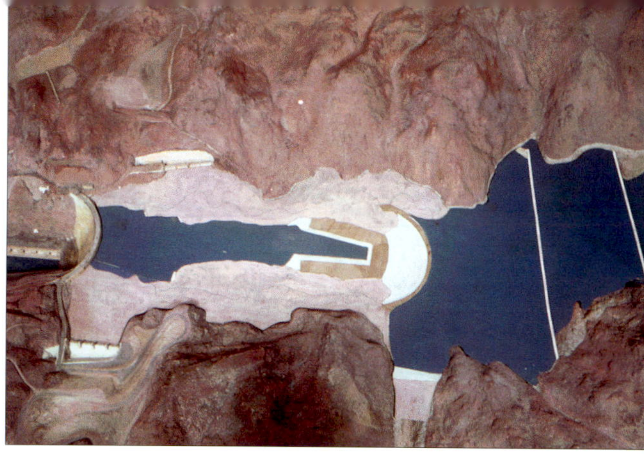

A World War II model of Hoover Dam, showing the camouflage system designed to protect the structure from enemy attack. Boulder Canyon Project, Arizona, Nevada. Date: October 15, 1998. Photographer: Andrew Pernick.

Mr. and Mrs. Tyrone Power visit Hoover Dam. With them are Mr. M. McKeen, who flew the Powers to Boulder City, and R.R. Spearman, who conducted the party through the powerhouse. Boulder Canyon Project, Arizona, Nevada. Date: May 26, 1939. Photographer: Cliff Segerblom.

THE WAR EFFORT IN THE PACIFIC NORTHWEST

The USAF Fighter Weapons School mission was to train pilots as weapons instructors. It has provided graduate training for nearly 50 years. Nellis Air Force Base, Nevada. Date: Unknown. Photograph courtesy of the United States Air Force.

Las Vegas Air Force Base's new mission after reactivation was the training of pilots in advanced single engine aircraft. Nellis Air Force Base, Nevada. Date: Unknown. Photograph courtesy of the United States Air Force.

While the film and aircraft industries showed their support for the allies, Reclamation and other United States industries prepared for the inevitable entrance into the War. Congress' 1940 Defense Appropriation Bill awarded enormous contracts to companies in the states of Washington and Oregon. The BPA provided eighty percent of the energy produced at the Corps' Bonneville Dam and Reclamation's Grand Coulee Dam to private wartime industries like the Aluminum Company of America, Reynolds Metal Company, Oregon Shipbuilding Company, Pacific Carbide and Alloys Company (then the largest of its kind west of the Mississippi), and the Pennsylvania Salt Company.[107]

By 1944, Reclamation quadrupled its hydroelectric power output. During the war, Reclamation powerplants produced forty-seven billion kilowatt hours of electricity to make 69,000 airplanes, 79,000 machine guns, 5,000 ships, seven million aircraft bombs, 5,000 tanks, and thirty-one million shells. Secretary Ickes boasted that, "The greatest expansion of electric power generating facilities ever made by a single agency in a single year was the wartime record established by the Bureau of Reclamation in the fiscal year 1944 to meet emergency industrial demands."

One of the most profound examples of the push to expand Reclamation's power during the war was the War Production Board's instruction in 1942 to temporarily move two 75,000 kW generators waiting for installation at Shasta Dam in California to the unfinished Grand Coulee plant in Washington to provide the necessary energy to industries in the Northwest.[108] This transfer required special adaptations of the generator pits by rerouting the penstocks to receive the smaller Shasta units that Reclamation originally designed to rotate in the opposite direction as those at Grand Coulee.

By 1945, Reclamation's production of nearly fourteen billion kilowatt hours made it the largest single power-producing bureau in the world. Reclamation was one of three bureaus, including the Corps and TVA, that had responsibility for planning and building Federal powerplants. By the end of World War II, Secretary Ickes boasted that, according to the most recent figures available, Interior hydropower facilities generated almost three times as much hydroelectricity as did all of South America in 1935. Ickes also expressed concern that the enormous power surplus developed during the war be rechanneled into constructive industrial and economic benefits during peacetime.[109]

Reclamation's supply of low-cost electricity in the Northwest attracted large defense industries to the region during and after the war. Shipyards, steel mills, chemical companies, oil refineries, and automotive and aircraft factories all required vast amounts of electrical energy that Reclamation could provide. Electricity from Grand

Coulee helped Boeing Industries produce airplanes during the war. Grand Coulee energy processed the aluminum for more than half of America's planes in World War II. Reclamation project power also processed food, lit military posts, and still continued to meet the needs of the civilian population in many areas.

The Columbia Basin Project also provided the impetus for the Federal government's decision to site the Hanford Engineer Works, a top secret atomic energy installation, in Hanford, Washington, in 1942.[110] The Hanford location allowed the installation to use the hydropower produced at Grand Coulee to run its plutonium-producing reactors. At the same time, the unsettled lands withdrawn for Reclamation's Columbia Basin Project provided the area with the necessary buffer zone between the reactors and human habitation. The Columbia River site also supplied sufficient cold river water to cool Hanford's reactors.

Beginning in January of 1943, a large block of electricity began to flow from Grand Coulee into a seemingly isolated area in southern Washington. To all appearances, the electricity simply disappeared into the desert, but in reality, it powered the reactors at Hanford. At the time, due to the classified nature of the project, BPA considered the unspecified amount of power Hanford received a "mystery" load.

On February 2, 1945, the first plutonium sample from Hanford reached Los Alamos, New Mexico, site of yet another portion of the far-flung Manhattan Project, which developed the hydrogen bomb. Hanford's plutonium eventually energized the atomic bomb dropped on Nagasaki, Japan, in August of 1945, effectively ending World War II.[111]

The demand for electricity created by the growing aluminum industry and the Hanford site caused a strain on the existing power network in the Northwest and prompted BPA to recommend building two or more of the dams proposed in one of the Corps' comprehensive river development plans, called 308 Reports. BPA also called for expansion of capacity at existing projects and expeditious authorization and appropriations for Reclamation construction activities at Hungry Horse on the Flathead River in Montana, a distant tributary of the Columbia River. Hungry Horse Dam would help regulate the flows of the Columbia, allow for more consistent water supplies, and increase power production at both Bonneville and Grand Coulee.[112] Congress authorized Hungry Horse Dam as a water and power project in 1944, but construction activities did not start until after the conclusion of the war.

Aerial view of a Hanford's 300 Area, where a variety of scientific, engineering, and technological activities take place. Date: Unknown. Photograph courtesy of the Department of Energy.

A view of Grand Coulee Dam, during a water release for the annual "Over The Dam Fun Run." Columbia Basin Project, Washington. Date: August 21, 1993. Photographer: David Walsh.

Footnotes for Chapter 6: Reclamation Power in World War II

[91] See Charles Keller, *The Power Situation During the War* (Washington, D.C.: Government Printing Office, 1921).

[92] Philip J. Funigiello, "Kilowatts for Defense: The New Deal and the Coming of the Second World War," *The Journal of American History* 56:3 (December 1969), 604-620.

[93] United States Department of the Interior, Harold L. Ickes, Secretary, "The War Program of the Department of the Interior," United States Department of the Interior, The National Archives Library, Washington, D.C., c. 1942.

[94] Harold Ickes, *Annual Report of the Secretary of the Interior, Post War Frontiers Edition* (Washington, D.C.: Government Printing Office, 1944), i-xiii.

[95] *Hoover Dam*, 4-5.

[96] The school was later renamed the Las Vegas Army Air Field. It was closed in January of 1947. In 1948, the base was reactivated as Las Vegas Air Force Base. A year later, it was renamed Nellis Air Force Base, which it remains today. Nellis Air Force Base, Early History, <http://www.nellis.af.mil/history/default.htm> accessed October 26, 2000.

[97] Myers, 197-198; *Hoover Dam*, 38-39; Stevens, 260-261.

[98] Thomas Schatz, *The Genius of the System: Hollywood Filmmaking in the Studio Era* (New York: Pantheon Books, 1988), 66.

[99] Robert Gottleib and Margaret FitzSimmons, *Thirst For Growth: Water Agencies as Hidden Government in California* (Tucson: University of Arizona Press, 1991), 51-53.

[100] Myers, 193.

[101] Stevens, 260.

[102] Established by the Motion Picture Producers and Distributors of America Production Code Administration, the Hollywood Production Code (also Production Code, or Hays Code) defined what could and could not be shown on film. Taboo subjects included brutality, sexual promiscuity, and the positive portrayal of illegal or immoral life. Originally drafted in 1930, Hollywood did not actually enforce the code until 1933. Hollywood officially eliminated the Hays Code in 1968 in favor of the motion picture rating system. Schatz, 167; Gerald Mast, *A Short History of the Movies*, Fourth Edition (New York: Macmillan Publishing Company, 1986), 226, 289.

[103] Charnan Simon, *Hollywood at War: The Motion Picture Industry and World War II* (New York: Franklin Watts, 1995), 16-19.

[104] Simon, 26-30.

[105] Simon, 47.

[106] Stevens, 262.

[107] Vernon M. Murray, "Grand Coulee and Bonneville Power in the National War Effort," *The Journal of Land & Public Utility Economics* 18 (1942): 134-139. An Executive Order temporarily suspended the preference clause in the Bonneville Act to first sell its power to public agencies like municipalities and cooperatives.

[108] <www.usbr.gov/power/edu/history.htm> accessed July 1, 1998; United States Department of the Interior, Bureau of Reclamation, *Hydropower: Water at Work* (Washington D.C.: Government Printing Office, c. 1981), 26; Department of the Interior, *Annual Report* (1944), 6-7; Paul C. Pitzer, *Grand Coulee: Harnessing a Dream* (Pullman, Washington: Washington State University Press, 1994), 252; Springer, 52-53.

[109] DOI, *Annual Report* (1944), v; DOI, *Annual Report* (1945), xiv-xv.

[110] General Leslie Groves, head of the Manhattan Engineer District (more popularly known as the Manhattan Project) made the final site selection for the plutonium-producing reactors. The Manhattan Project, under the jurisdiction of the Army Corps of Engineers, was responsible for development and construction of the first atomic weapons. The project brought together researchers and scientists in remote locales, including Los Alamos, New Mexico; Oak Ridge, Tennessee; and Hanford, Washington. The Oak Ridge facility produced uranium 235, while the Hanford facility produced plutonium 239, and the Los Alamos facility constructed the trigger assembly necessary for atomic weapons.

[111] Bonnie Baack Pendergrass, *Public Power, Politics, and Technology in the Eisenhower and Kennedy Years: The Hanford Dual-Purpose Reactor Controversy, 1956-1962* (New York: Arno Press, 1979), 45; White, 52. Also see Michele Stenehjem Berber, *On the Home Front: The Cold War Legacy of the Hanford Nuclear Site* (Lincoln: University of Nebraska Press, 1997).

[112] Robert E. Ficken, "Grand Coulee and Hanford: The Atomic Bomb and the Development of the Columbia River," in *The Atomic West*, Bruce Hevly and John M. Findlay, eds. (Seattle: University of Washington Press, 1998), 27.

Mead Substation 230 kV Type Pk-4b circuit breakers. Boulder Canyon Project, Arizona, Nevada. Date: Unknown. Photographer: Unknown.

The USAF Thunderbirds fly in formation over Hoover Dam. Date: Unknown. Photograph courtesy of the United States Air Force.

Aerial view of Hanford's 300 Area, which is largely geared to advanced environmental and nuclear research and development. Date: Unknown. Photograph courtesy of the Department of Energy.

CHAPTER SEVEN

Throughout the West, decommissioning of wartime industries created a significant block of surplus capacity, much of which was available from Reclamation's projects. This surplus energy capacity opened new opportunities for further western municipal and industrial development. Many wartime industries, specifically the ones in southern California, simply altered their operations and continued to do business as usual.

POSTWAR DEVELOPMENT IN THE PACIFIC NORTHWEST

Initially, the end of the war brought few changes to defense industries in the Pacific Northwest – though early indications predicted otherwise. The majority of the wartime industries simply adapted to peacetime demands and changed their operations accordingly. For example, instead of being decommissioned, the region's wartime aluminum industry reorganized under Federal auspices and agreed to use forty percent of the energy generated at Bonneville and Grand Coulee.[113] This reorganization, and others like it, maintained the industrial demand for electricity in the Pacific Northwest.

As the region and its industries adjusted to a postwar economy, Reclamation returned Grand Coulee's borrowed generators to Shasta, finishing its construction, and finally completed Grand Coulee's first two powerhouses in 1951. At that time, powerplants at Grand Coulee had the largest installed capacity of any power station in the world.

As part of its service to the public market, BPA promoted electricity at low cost, distributed power, and created new markets. Federal wartime investment for national defense in the Pacific Northwest gave the region an expanded industrial economic base for peacetime. Eventually, so many industries in the Pacific Northwest used Federal hydropower that periodic low flows on the Columbia River often disrupted manufacturing in the region.[114]

The Hanford Engineer Works (the secret wartime "bomb factory") also successfully weathered the change from a wartime to a peacetime economy. While the closure of other defense plants led to unemployment and subsequent concern about generating enough civilian demand for electricity, Hanford continued to function as part of a system of laboratories, manufacturing units, and testing grounds across the country.[115] During World War II and the subsequent Cold War, Hanford became a massive, isolated defense reservation for the production of plutonium and, eventually, nuclear power. Grand Coulee generators provided electricity to three plutonium-producing reactors during World War II; this number expanded by five during the early part of the Cold War, for a total of eight operating reactors at the facility in 1955.

Grand Coulee Dam, view showing east side section after its final cleanup. Columbia Basin Project, Washington. Date: November 1937. Photographer: Unknown.

Bolts and nuts, weighing 193 lbs, were used to couple the sections of the shafts for the two 75,000 kW generators installed in Grand Coulee Dam left powerhouse. These were the units originally intended for Shasta but moved to Coulee due to electrical demand during the war. Columbia Basin Project, Washington. Date: December 29, 1942. Photographer: Unknown.

President Lyndon B. Johnson signed legislation authorizing construction of the third powerplant at Grand Coulee Dam on June 13, 1966. Standing before a large artist's drawing of the third powerplant and Grand Coulee Dam left to right: Congressman Thomas S. Foley, Senator Warren G. Magnuson, and Senator Henry M. Jackson, all of Washington state; President Lyndon B. Johnson; Vice-President Hubert H. Humphrey; and Secretary of the Interior Stewart L. Udall. Columbia Basin Project: Third Powerplant, Washington. Date: June 13, 1966. Photographer: Jim Aycock.

General interior view of Shasta Powerplant. Central Valley Project: Kennett Division, California. Date: February 17, 1944. Photographer: P.E. Norine.

During the Presidential administrations of Eisenhower (1953-1961) and Kennedy (1961-1963), the issue of installing a dual-purpose reactor[116] to produce both plutonium and electricity at Hanford attracted the attention of both public and private interests who wanted to participate in the control and development of the atomic bomb and nuclear energy. In 1961, President Kennedy partially resolved the issue by requesting sixty million dollars to build a generating plant that would use waste heat (steam) from a plutonium reactor under construction in the weapons program at Hanford, Washington, to generate electricity.[117]

Kennedy's request did not completely end discussions regarding the development of atomic energy. In fact, another two years of negotiation ensued about who would develop and construct the facility and then market the generated energy. Eventually, a compromise was reached with private interests, and the Washington Public Power Supply System (WPPSS – pronounced "Whoops" in Washington State) procured the right to develop the second part of the Hanford dual-purpose reactor. WPPSS, a consortium of utilities, would lease a small section of land near the reactor from the Atomic Energy Commission (AEC) for constructing generating facilities. The AEC would sell the steam produced in the reactor to WPPSS. WPPSS would use the steam to produce electricity it would sell to BPA. In return, BPA would supply WPPSS and participating agencies with an amount of firm power equal in value under BPA rates to the annual operating and capital costs of the Hanford generating facilities; essentially, the two utilities were exchanging energy (hydro for steam).

Even prior to the decision on how to, and who would, generate power at Hanford, construction of the ninth reactor, this one a dual-purpose reactor capable of producing both plutonium and electricity, began in 1958. In October of 1963, WPPSS began work on the generating facilities. Construction was finally finished in 1966, when the utility synchronized the first generator with the Northwest Power Pool, signaling the official start of commercial power production at the facility.

What originally looked like a successful and profitable compromise between BPA and WPPSS never quite achieved its projected status. The Hanford reactor proved unreliable, so steam was often unavailable for sale. This forced BPA to meet its exchange obligations without benefit of receiving additional energy from Hanford. To meet these needs, BPA often had to import high-cost energy from California over the newly constructed Pacific Northwest-Pacific Southwest Intertie.[118]

Today, with the addition of the Third Powerplant, Grand Coulee has the capacity to produce more energy than any other powerplant in the United States and is the third largest hydropower plant in the world. According to the Energy Information Administration, as of January 1, 1998, hydroelectricity provided eighty-five percent of the electric capacity in the Pacific Northwest (Washington, Oregon, and Idaho). Reclamation generated twenty-three percent of the hydroelectricity in the region, which made it the largest electric utility in Washington and the fourth largest in Idaho.

The Columbia River hydroelectric projects actually helped define the Pacific Northwest region. Electricity offered the opportunity for industrial development. BPA, the Federal marketing and transmission bureau for hydropower, defined the region's boundaries through its network of transmission lines. Prompted by wartime power needs, BPA created a lasting "Northwest power pool" of interconnections among both public and private utilities, transmission, and generation facilities.[119] These interconnections paved the way for future cooperation between public and private interests and provided a model for future regional interconnections and power grids throughout the United States.

POSTWAR SOCIAL TRANSFORMATION

The Pacific Northwest was not alone in its postwar transformation. As a region, the American West emerged from World War II as a very different place. Population exploded, and modern new cities boomed. These activities symbolized progress, a bright economic future, and a new urban landscape for the West. With available electrical power no longer needed for war purposes, western towns could add industry to agriculture and mining as a mainstay of their local economies. Reclamation powerplants enlarged for wartime use could support these new economies by developing additional sources of low-cost electrical energy for recently created and expanding communities and their accompanying commercial and industrial developments.[120]

The enormity of the war and its pervasive influence on the home front changed people's hopes and expectations about their lives and their country. Popular culture in the form of movies, advertisements, and comic books helped shape a national experience of renewed optimism and purpose. In addition, the war significantly changed domestic life, and earlier promises regarding technology's ability to improve the standards of comfort, convenience, and cleanliness came to fruition in the postwar period. In 1940, about one-third of Americans still pumped water from their own wells, two-thirds had no central heating, less than half had indoor plumbing, and one-third still cooked with wood and coal.

The revitalized postwar economy allowed wider access to affluence, and affluent spending habits brought appliances like toilets, refrigerators, air conditioners, and washing machines into many homes. The new emphasis on maintaining single-family homes contributed to the rise in domestic electrical loads and a new suburban market for electric energy.[121]

Spud Indoors – couple eating dinner with "Idaho Russ." Date: February 17, 1967. Photographer: Jack Williams. Photo courtesy of Idaho State Historical Society, 1996.26.585.

As a result of the rewinding of Davis' Unit 2 generator, water was bypassed through the spillways of the dam. Parker-Davis Project, Arizona, California. Date: April 12, 2000. Photographer: Andrew Pernick.

At first, manufacturers found it difficult to interest the public in climate control for their home. Air conditioning's then more popular function was to control humidity in factories. Industry marketers promoted air conditioning to utilities as a way to sell electricity during the summer, when demand dropped because wintertime heating units went unused. New postwar-style construction methods, encouraged by the modernization efforts of President Truman's "Fair Deal" equal housing programs, lent support to the adoption of air conditioning in urban residential areas.

Consumer acceptance of air conditioning, as well as other electrical appliances, created an enormous increase in the Nation's domestic demand for energy. Although gaining consumer approval proved lengthy and difficult, once people adopted air conditioning – as its cost and that of many appliances fell – other values and attitudes toward comfort and living changed. In some areas, an air conditioned home came to be considered a necessity rather than a luxury. This was especially true in the hot and dry regions of the West. The demand for energy-consuming devices, such as air conditioners, grew so fast that both public and private sectors of the electrical industry had a difficult time meeting the demand, particularly in the western states.[122]

Most scholars agree that electrical modernization in many parts of the United States stagnated prior to World War II because private electric utilities maintained high rates.[123] After the war ended, rates fell and electrical usage increased. The years between 1945 and 1965 became the "Golden Age" for electric utilities, with exponential load growth and continual cost reductions. The government and industry saw great advantages in system interconnections among the many electricity generators to reduce capital and fuel costs.[124] No longer simply a commodity, a spectacle, or a byproduct of irrigation, electricity evolved into an expected right.

FEDERAL EXPANSION AND THE REVIVAL OF PUBLIC POWER

For the most part, planning for major new dams slowed in the postwar years, but ongoing and previously authorized construction activities continued. In the Columbia River system, Federal entities completed dams previously authorized by Congress at a rate of one to three per year between 1952 and 1958.[125] The government continued to operate its projects. Although Reclamation maintained irrigation as the primary project purpose, electrical power remained key to postwar progress. The 1944 Flood Control Act furthered public power responsibilities by authorizing more hydroelectric dams, including those in the Pick-Sloan Missouri River Basin Program (PSMBP), which established full river basin development for the Missouri River as a Federal activity. After the war, Reclamation also resumed its construction of projects and transmission systems stalled by the war, such as Davis Dam and

Powerplant, a water storage and power project located downstream from Hoover Dam.

When Harry S. Truman became President in 1945, public and private utility interests still debated the role of the Federal government in power generation. Various bureaus, including Reclamation and the Corps, were responsible for Federal energy production and public power policy. New Dealers retained their political influence, and the opportunities for nuclear power and power released from war production reignited the old fights over public and private power that had stalled during the wartime emergency.[126] This time, the debate had a new facet as nuclear power became a point of contention. Both the scientists, who developed nuclear energy, and the military, for whom they developed it, challenged Congress for oversight of the new, powerful, and little understood technology. The question of nuclear power also saw a resurgence of private power influence. A civilian commission, in cooperation with joint congressional committees, developed legislation for the Atomic Energy Act of 1946.[127]

That year, Secretary of the Interior Harold Ickes issued a memorandum to his staff reiterating that Federal dams should include generation facilities, where feasible, and that these facilities should grant preference in power sales to public bodies and cooperatives. Ickes also determined that "power disposal shall be such as to encourage widespread use and prevent monopolization."[128]

The Truman administration presided over the most extensive Federal power development period in history, both in generating capacity and transmission line expansion, a notable achievement for that period. The Corps and Reclamation built more structures related to power during this period than at any other time. The government faced a large national wartime debt, and President Truman spent much of his time on international issues, such as foreign aid for war-torn countries in Europe. At the same time, with the massive development of Federal power during the war now over, private power again began to assert its interests. The public versus private power issue re-emerged as a key campaign issue in President Truman's re-election, one that symbolized progress, development, and advancement.[129] President Truman gradually developed a public power policy program to expand upon the New Deal legislation and pushed Congress for large power appropriations.[130]

Following the war, private power interests resumed their fight against the government's control of the enormous energy production at water projects. They opposed the creation of new regional authorities modeled after the TVA, to sell Federal power to "preferred customers," and the jurisdiction of the Federal government over the construction and operation of many hydroelectric projects. To make matters more difficult for public power, a conservative Congress was less supportive of publicly owned utilities. Within the new Cold War climate, when the nuclear arms race made Russia and Communism of any form

Construction of Shoshone (Buffalo Bill) Dam. Shoshone Project, Wyoming. Date: Unknown. Photographer: Unknown.

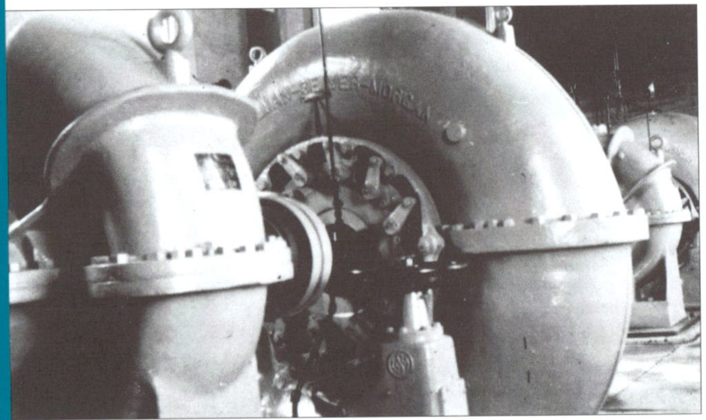

Interior view of Thomas Point Pumping Plant, showing machinery. Lower Yellowstone Project, Montana, North Dakota. Date: Unknown. Photographer: Unknown.

Erecting the surge tank, Guernsey Dam Powerhouse. North Platte Project, Wyoming. Date: May 2, 1927. Photographer: Unknown.

the enemy, politicians such as Senator Joseph McCarthy accused numerous individuals and industries of "un-American" activities during a period known today as the "Red Scare." McCarthy's sympathizers labeled both Roosevelt's and Truman's New and Fair Deal policies socialistic.

The 80th Congress, from 1947-1949, was the first Republican Congress in eighteen years. However, while criticizing Federal spending, Congress still granted most of Reclamation's requested appropriations, awarding them funds in record amounts. President Truman and others cited frequent floods in 1947 to gain public support for Federal dam building and water control. Reclamation Commissioner and New Deal proponent Michael Straus instructed his regional directors to think of themselves as social planners and to concentrate on "distributing the benefits of natural resource development as widely as possible." Commissioner Straus launched a public relations campaign to emphasize the technological promise and prosperity of the West to the public. Reclamation used many of these appropriations to increase its generating capacity at existing projects. The 1948 election brought in a new Congress, which promised Reclamation money for more powerplants.[131]

The Korean War also stepped up Federal power needs and prompted additions to transmission systems. The government built 13,000 miles of transmission lines during President Truman's administration versus the 11,000 miles built during all previous administrations.[132] Still, although the development of Federal power progressed, private industry retained the dominant position it had achieved prior to and during World War II. During this time, the National Association of Electric Companies emerged as one of the most powerful of many private utility lobby groups.

President Truman's successor was less enthusiastic about the Federal government's role in electrical power. When Dwight D. Eisenhower assumed the Presidency in 1953, he adopted a policy of dissuading Congressional authorization of any new multipurpose projects and promised to provide tax incentives for private development. In August of that year, Interior announced that the government would only finance large multipurpose projects that private enterprises could not fund. A policy that coordinated private and public power needs emerged out of the Eisenhower Presidency.[133]

Components of a forty-two inch needle valve for Stony Gorge Dam. Orland Project, California. Date: June 26, 1928. Photographer: Unknown.

Of more than 180 Reclamation projects, Congress authorized only seventy prior to World War II. It approved the remainder during and after the war in both small authorizations and others of great importance to Reclamation's power program. These included the Pick-Sloan Missouri Basin Program in the Flood Control Act of 1944, the Colorado River Storage Project Act (1956), and the Third Powerplant at Grand Coulee Dam on the Columbia Basin Project (1966).

Boysen Dam construction. Pick-Sloan Missouri Basin Program: Boysen Unit, Wyoming. Date: 1947-1952. Photographer: Unknown.

Section of a turbine. Salt River Project, Arizona. Date: April 16, 1906. Photographer: J.M.M.

Performing diagnostic tests on a hydroelectric generator at Little Harquahala pumping plant to evaluate the condition of the armature winding insulation. Central Arizona Project, Arizona. Date: March 1995. Photographer: Unknown.

Footnotes for Chapter 7: Reclamation Power in Peacetime

[113] White, 3; Ficken, *Rufus Woods*, 217.

[114] Pitzer, *Grand Coulee*, 253; William E. Riebsame, ed., Hannah Gosnell, David Theobald, *Atlas of the New West: Portrait of a Changing Region* (New York: W.W. Norton and Company, 1997), 86.

[115] White, 82-84.

[116] A dual-purpose reactor is used both in the development of atomic materials – either plutonium in the case of Hanford or uranium – and the generation of electricity. The production portion of the reactor facility produces plutonium 239. Then, instead of being released into the air or diffused, the resulting "waste heat" is transported via a system of pipes to a separate generating facility which operates much like a traditional steam generating plant.

[117] Rudolph and Ridley, 119. Also see Pendergrass.

[118] Opponents of the dual-purpose reactor at Hanford originally proposed an interconnection between the Pacific Northwest and California in lieu of the generating facilities. President Kennedy, however, supported both ideas and called for construction of generating facilities at Hanford and a Pacific Northwest-Pacific Southwest Intertie. Construction on both programs occurred during the 1960s. For more information on the Intertie, please see the later section on "Reclamation and the Interconnection of Western Power Systems." Pendergrass, 131, 132, 176, 177.

[119] White, 64-65, 70.

[120] "Natural Resource Problems," Annual Report (1946), 77-79; Paul Pitzer, "The Mystique of Grand Coulee Dam and the Reality of the Columbia Basin Project," Paper, Western History Association, Tacoma, Washington (October 14, 1989), 5.

[121] Ruth Schwartz Cowan, *More Work for Mother: The Ironies of Household Technology from the Open Hearth to the Microwave* (New York: Basic Books, 1983); Gail Cooper, *Air Conditioning America: Engineers and the Controlled Environment, 1900-1960* (Baltimore: Johns Hopkins University Press), 1998; Mark H. Rose, *Cities of Light and Heat: Domesticating Gas and Electricity in Urban America* (University Park, Pennsylvania: Pennsylvania State University Press, 1995), xv.

[122] See Cooper; Harold Ickes, *Annual Report of the Secretary of the Interior*, Natural Resource Programs (Washington D.C.: Government Printing Office, 1947), 38-39.

[123] See Tobey; White, 71-73; John A. Casazza, *The Development of Electric Power Transmission* (New York: IEEE, 1993), 27.

[124] The process of interconnection receives more detailed attention later in this study. Casazza, 27; Hoy, Robinson, and Armstrong, 312.

[125] White, 75.

[126] Rudolph and Ridley, 87.

[127] Rudolph and Ridley, 91.

[128] Phyllis Komarek de Luna, *Public Versus Private Power During the Truman Administration* (New York and Bern: Peter Lang, 1997), 14.

[129] Waltrip, 2.

[130] Waltrip, 130-135; Komarek de Luna, 169-170; David Arlin Kathka, "The Bureau of Reclamation in the Truman Administration: Personnel, Politics, and Policy" (Ph.D. dissertation, University of Missouri, 1976), 27.

[131] Waltrip, 67-69.

[132] Waltrip, 109, 130-31.

[133] Springer, 59; Komarek de Luna, 172-173.

"Yellowtail Switchyard." Pick-Sloan Missouri Basin Program: Yellowtail Unit, Montana, Wyoming. Date: Unknown. Artist: Michael Frary.

CHAPTER EIGHT

As the electric power industry developed, so did the need for interconnected transmission lines. Soon after the initial development of electrical systems at the turn of the twentieth century, engineers, operators, and politicians realized the need for coordination of multiple electric systems, as well as individual power lines. During the 1920s, integrating these systems became part of the public versus private power debate. Advocates of private development initiated the concept of "Superpower," the interconnection of private utility systems. Advocates of public power christened their idea for the interconnection of regional systems into a national network, "Giant Power." However, the argument between Superpower and Giant Power was not resolved before the country's attention turned to other immediate issues, leaving public and private interconnections to develop without a national policy.

Individual private companies connected their own facilities as they grew, creating self-contained miniature grid systems that traversed the United States. Prior to the Depression, private utilities connected systems by using holding companies to consolidate various utilities into centralized entities. These grids connected different types of generating facilities – steam, hydropower, coal, and, later, even nuclear. For the most part, a utility created a monopoly within a given area, making sure that their transmission lines were the only ones capable of providing electricity to residents and businesses within its boundaries. As private utilities grew, so did their grid systems. Very often, growth took the easy form of buying out other utility systems and incorporating the new systems into an existing one. These large-scale grids allowed each utility to supply power to its customers from various points within their system, making the entire transmission system more efficient.

Federal and state governments also integrated their own transmission systems as their influence in the power industry grew. Interconnection of the government systems did not occur as easily as those of private utilities. The logistics of interconnection proved very difficult, largely due to the wide-ranging and often isolated locations of many Federal projects. As construction of Federal water and power projects increased, a Federal grid system emerged.

THE BONNEVILLE POWER ADMINISTRATION

Regulation of the power industry began in 1935, when the Federal government enacted the Public Utility Holding Company Act to address the problem of holding companies in the utility industry. This act required that utility holding company systems integrate and simplify into public utility systems. Although the requirement did not directly apply to Federal utilities, the United States began condensing its multifaceted and multi-agency holdings into single unified entities. The first truly integrated Federal

The double circuit cantilever rim towers on the Arizona rim at Hoover Dam. Boulder Canyon Project, Arizona, Nevada. Date: May 2, 1951. Photographer: Mark Swain.

Bell Type 116 inch turbine runner (Unit No. 1) at Siphon Drop Powerplant. Yuma Project, Arizona. Date: June 21, 1926. Photographer: R.G.H.

system included hydropower developments of the Corps and Reclamation along the Columbia River. To better manage the two bureaus' projects together, Congress created the first Federal marketing bureau, the Bonneville Power Administration.

Abandoned farm near Lincoln, Nebraska. Date: October 1938. Photographer: Unknown.

The Great Depression and President Roosevelt's New Deal brought new opportunities for Reclamation and the Corps to construct dam and hydroelectric facilities throughout the United States. The Corps continued to construct facilities in the West as part of New Deal public works activities, including Fort Peck Dam in Montana and Bonneville Dam on the Columbia River on the border between Oregon and Washington.[134] At the same time that the Corps began construction on Bonneville Dam, Reclamation began work on Grand Coulee Dam on the Columbia River in eastern Washington. Both dams contained hydroelectric facilities, and since they both served the same area, the two bureaus had to reach some sort of compromise regarding power distribution.

As construction of Bonneville Dam neared completion, discussions began about who should distribute the energy from the large hydroelectric dams of two different government entities in the same region. Interested in region-wide distribution, Senator James J. Pope of Idaho introduced a bill in 1935 that would create the Columbia Basin Authority, modeled after the TVA. Both the Corps and Reclamation opposed this idea since they thought it denied them their right to operate and control the powerplants they constructed. Formation of a Columbia Basin Authority would separate the construction portion of a project from its operation – two activities that went hand-in-hand, according to both bureaus. Another bill that year suggested that each government bureau distribute energy from its own project with the Federal Power Administration setting the rates. This option would allow the sale of inexpensive power to major industries. President Roosevelt and Senator Pope were concerned that this would fail to provide widespread electrical service to expanding rural and domestic markets that were underserved by existing utilities. Many opponents also argued that neither bureau had the qualifications or the mandate to successfully administer projects with the power production capability of Bonneville and Grand Coulee.

In 1936, the Pacific Northwest Regional Planning Commission recommended that a central grid unite the two power systems through uniform transmission lines. An independent corporation would operate the system. The Secretary of War, concerned that marketing and transmission duties would be taken away from the Corps, rejected the proposal. Roosevelt then appointed a special committee, chaired by Secretary of the Interior Ickes, to further investigate the idea. Instead of an independent corporation, the committee decided to form a government bureau, under Interior, to ensure low energy rates and assume the

First bucket of concrete placed at Angostura Dam. Pick-Sloan Missouri Basin Program: Angostura Unit, South Dakota. Date: 1947. Photographer: Unknown.

additional responsibility of marketing energy generated at Bonneville and Grand Coulee Dams. The new bureau would also market energy from further government hydroelectric developments in the Columbia Basin, giving at least the semblance of coordinated regional development for the region. The new entity, approved as part of the 1937 Bonneville Project Act, eventually became known as the Bonneville Power Administration.[135] In 1977, BPA was transferred to the Department of Energy, where it continues to function today.

THE PICK-SLOAN MISSOURI BASIN PROGRAM[136]

The Missouri River originates where the Gallatin, Madison, and Jefferson Rivers join together near the town of Three Forks, Montana. The mainstem twists and turns for nearly 2,500 miles before joining the Mississippi River near St. Louis. Tributaries of varying length feed the river along its 2,500 mile course. The Missouri River basin includes all of Nebraska; the majority of Montana, North Dakota, South Dakota, Kansas, and Wyoming; as well as parts of Colorado, Iowa, Minnesota, and Missouri, and a small portion of Canada. The differing needs within this area fuel conflicts surrounding the development of water resources in the basin.[137]

The success of the TVA and BPA prompted President Roosevelt to recommend additional comprehensive river basin developments, including power, throughout the United States. The Missouri River Basin, as yet relatively undeveloped, offered an ideal proving ground for Roosevelt's ideas. One of the largest river basins in the United States, the Missouri River Basin covers 528,000 square miles of drainage area, a land area equal to one-sixth of the continental United States.

The greatest problem with Roosevelt's vision of development of the basin was the question of who would implement it and what benefits would be provided. In the 1930s, both the Corps and Reclamation developed plans for large-scale development of the Missouri River Basin. The Corps focused primarily on large-scale flood control, navigation, and power generation to meet the needs of the Lower Basin States. Reclamation focused on irrigation to meet the needs of the Upper Basin States. Roosevelt originally envisioned an organization similar to the existing TVA, a Missouri Valley Authority (MVA), to coordinate development. However, the governors of the ten states, among others, were

A transmission pole, burned off by lightning. Buford-Trenton Project, North Dakota. Date: July 27, 1909. Photographer: W.H.M.

strongly opposed to the MVA idea. None of the states (or even the Corps and Reclamation for that matter) wanted to relinquish ultimate control of the river – or any structures that might be built on the river – to a single Federal entity. Similar proposals in 1937 for a Central Valley Authority in California and an Arkansas Valley Authority (AVA) in 1941 met opposition in Congress. The MVA bill did not pass either, and the contest between the Corps and Reclamation for development of the river quietly continued. As each tried to gain control of the river basin, both bureaus wrote their own proposals for development.

Aerial view of Glendo Dam showing reservoir on the right and powerplant at center left. Pick-Sloan Missouri Basin Program: Glendo Unit, Wyoming. Date: June 1983. Photographer: Jim Todd.

The United States' entry into World War II in 1941, and a partial recovery from the Great Depression, turned national attention away from large-scale national public works and river basin development. In 1943, however, a severe flood along the Missouri River renewed interest in development of the river valley. At the same time, the Missouri River Basin States – and, to a certain extent, the Nation as a whole – began to address the issue of a return to a peacetime economy; although the war was not yet over, an eventual victory seemed fairly certain. Taking a proactive stance, the states attempted to initiate several federally funded construction projects to put returning soldiers to work. Many of their proposed construction activities took the form of dams and irrigation works, which also provided homesteads and farming opportunities for returning veterans.[138]

On May 13, 1943, following the floods on the Missouri River, the House Flood Control Committee ordered the Corps to study flood control needs in the basin and prepare a plan of action. The task of preparing a report to Congress fell upon Colonel Lewis A. Pick, the Missouri Basin Division Engineer in Omaha, Nebraska. Even though he had minimal experience in flood control construction, Colonel Pick wasted little time preparing his proposal, largely based on the earlier basin study prepared by the Corps. Roughly ninety days later, he submitted a ten-page report, which became known as the Pick Plan, to the Chief of Engineers in Washington, D.C.

The Pick Plan called for the Corps to construct three groups of projects. The first portion of the project plan called for construction of 1,500 miles of protective levees from Sioux City, Iowa, to the Mississippi River to help alleviate future flooding and aid navigation. The second phase proposed construction of eighteen tributary dams, eleven of which had been previously authorized. Five new tributary dams lay on the headwaters of the Republican River in the Lower Basin. The remaining two new tributary dams were high on the Yellowstone and Bighorn Rivers in Montana and Wyoming, respectively. The third and last portion of the project plan outlined a series of five huge multipurpose dams on the mainstem of the Missouri River above Sioux City. In total, the Pick Plan proposed construction of 1,500 miles of levees and twenty-three dams.

Although Pick's plan addressed many of the problems along the river, including the concerns of the Lower Basin States, the overall plan lacked details such as exact water flow estimates, reservoir specifications, and procedures for distribution of power and irrigation water. Pick also left out an analysis of project costs in relation to anticipated benefits, and failed to answer the overriding question of whether there would be enough water available for all intended purposes. Relying largely on his own public-relations experience, Pick formulated a set of proposals and presented them so that the ordinary citizen could easily comprehend the overall plan, which also served to facilitate Congressional review. His plan asserted the Corps' dominance in the river basin and directly challenged the rival Bureau of Reclamation.

In February of 1944, before the Pick Plan had been submitted to Congress or had undergone Executive review, Colonel Pick began promoting his plan throughout the basin on his own initiative. President Roosevelt countered by asking the Director of the Budget, currently reviewing the plan, to return it to the Secretary of War for revision, pointing out that the plan did not fit current Presidential objectives and conflicted with existing Reclamation plans in the Upper Basin. Roosevelt also criticized the Plan for failing to maximize power development. During the Depression and World War II, the Federal government had emerged as a major producer of hydroelectricity, and FDR wanted to ensure that this trend continued after the war. Meanwhile, however, the House Flood Control Committee, without awaiting further Executive action, began hearings on the Corps' Plan. Not surprisingly, governors of the Upper Basin states attending the meeting echoed the President's objections, seeing the plan as a threat to their own desire for irrigation development.

The hearings on the Pick Plan pitted the Upper and Lower Missouri Basin States against one another, much like the friction among the Upper and Lower Colorado River Basin states during the 1920s. The Lower Missouri River Basin States believed in the necessity of additional flood control along the river, while the Upper Basin States advocated construction of irrigation works to further develop the region. The two factions created a stalemate because neither side had enough power and influence with Congress to pass a bill addressing its own needs. At this juncture, the Bureau of Reclamation stepped into the fray with its own comprehensive plan for development of the basin.

Intending to formulate an all-inclusive development plan, Reclamation had previously conducted a five-year study of the region's water needs. Presentation of the Corps' Pick Plan to Congress prompted speedy completion of the Sloan Plan. Responsibility for completion of the plan fell to William Glenn Sloan, then Assistant Regional Director of the Upper Missouri Region in Billings, Montana, who had already done considerable work on the project.

Spilling from Kortes Dam during the record water year of 1983. Kortes Powerplant is seen at the foot of the dam. Pick-Sloan Missouri Basin Program: Kortes Unit, Wyoming. Date: June 23, 1983. Photographer: Jim Todd.

Canyon Ferry Dam. Pick-Sloan Missouri Basin Program: Canyon Ferry Unit, Montana. Date: 1974. Photographer: Unknown.

Fort Peck Dam under construction, Montana. Date: c. 1930s. Photograph courtesy of United States Army Corps of Engineers.

Reclamation submitted Sloan's 211-page Missouri Basin Report, which soon became known as the Sloan Plan, to Congress on May 4, 1944. Much more detailed than the Pick Plan, the Sloan Plan proposed ninety projects, emphasizing irrigation and power development rather than flood control and navigation. By this time, the House of Representatives had approved the Pick Plan.

Sloan's Plan called for roughly the same amount of total reservoir capacity as the Pick Plan; however, Sloan spread the storage out over a series of smaller dams located in the headwaters and on tributaries, as opposed to downstream and on the mainstem of the Missouri. Sloan eliminated three large Pick Plan projects on the Missouri and modified three Corps structures, including the newly constructed Fort Peck Dam, which adapted the original purpose of the project from flood control and navigation to irrigation and power. Two mainstem dams remained, and Sloan added another to permit construction of massive diversion systems to irrigate the areas of the Dakotas hardest hit by drought. Sloan did not completely overlook the flood control and navigation provisions in the Pick Plan; he incorporated the massive navigation channel proposed in the first phase of the Pick Plan and added water for municipal and domestic uses.

Similar to the Pick Plan, the Sloan Plan made no attempt to ensure adequate water supplies for all project purposes, though it did acknowledge that shortages might eliminate future navigation. Sloan, however, did provide economic justification for his plan in the form of Federal sales of irrigation water and hydroelectric energy, which he anticipated would more than cover all project costs.

With release of the Sloan Plan, the impasse between the two sides remained. Logically enough, the Upper Basin States, which were interested in irrigation, supported Reclamation and the Sloan Plan. The Lower Basin States, which were interested in flood control and navigation, favored the Pick Plan. With the battle lines in evidence, the *St. Louis Dispatch* offered another alternative: to establish a unified water development plan under the direction of a Missouri Valley Authority.

Though plans for an MVA failed when first presented to Congress in the 1930s, a Federal mandate for coordinated development of the region seemed like an ideal solution to the stalemate between the Corps and Reclamation. The news of renewed interest in an MVA caused both the Corps and Reclamation concern. Neither bureau liked the idea of turning control of the river basin over to a different Federal entity. Realizing that a compromise had to be reached and faced with growing public, Congressional, and Presidential support for the MVA, the Corps and Reclamation scheduled a conference in Omaha, Nebraska, to draft a joint engineering report.

The result of the two-day conference in 1944 was a one-page agreement that merged the two plans. Representatives of the Corps and Reclamation combined more than 1,400 pages of technical data, specified an exact number of projects, and determined their precise jurisdiction. With a few minor exceptions, each group merely accepted the other's proposed projects. Reclamation accepted the five new mainstem dams proposed by the Corps (Garrison Dam in North Dakota, and Oahe, Big Bend, Fort Randall, and Gavins Point Dams in South Dakota), while the Corps agreed to Reclamation's twenty-seven dams in the Yellowstone Basin. Both parties agreed on the improvement and incorporation of Fort Peck Dam. The two bureaus exchanged sites on the Niobrara, Platte, and Kansas Rivers.

The Pick-Sloan Plan, as finally developed, resulted in an extremely complex series of interactions between the Corps and Reclamation in which overlapping authorities occurred. Generally, however, responsibilities fell out this way. Overall, the Corps retained jurisdiction over the proposed mainstem dams, the key structures of the project, and the tributary projects primarily designed for flood control and navigation. Reclamation retained jurisdiction over the majority of tributary projects as well as irrigation development. The development of hydropower and other benefits fell to both bureaus at their respective sites.

Of the 113 projects proposed in the original two plans, 107 made it into the new Pick-Sloan Plan. When fully developed, the comprehensive plan, officially labeled the Missouri River Basin Development Program (later renamed the Pick-Sloan Missouri Basin Program), aimed to eliminate all flooding of any consequence on the Missouri and its tributaries and, at the same time, to provide irrigation water to more than five million acres in the Upper Basin. At the same time, the combined plan preserved future water development opportunities in the Upper Basin states. Once Reclamation and the Corps completed the joint plan, the two newly allied bureaus wasted little time presenting it to Congress, which received it favorably.

Not everyone reacted positively to the Pick-Sloan Plan. James S. Patton, President of the National Farmers Union, described the merger as "a shameless, loveless shotgun wedding."[139] He contended that the plan made no attempt to justify or consolidate costs, specify project dimensions, or remove project overlap. Patton and other detractors observed that neither bureau even tried to determine if enough water existed to provide for both irrigation and navigation, and that the new plan notably lacked any details

W.G. Sloan (far right and pointing) next to Colonel Pick and party at the future site of Yellowtail Dam near Fort Smith, Montana. A tour of the site used the suspension bridge to view the canyon. Pick-Sloan Missouri Basin Program. **Date: Unknown. Photographer: Unknown.**

concerning the development and distribution of hydroelectric power, and the problems of administrative coordination. Despite its opponents, the Pick-Sloan Missouri Basin Program passed through Congress with relative ease. The MVA supporters, while vocal in their dislike of the compromise plan, had neither an alternative proposal nor the political support necessary to block the legislation.

President Roosevelt maintained his faith in the necessity of a Missouri Valley Authority, even as he signed the Flood Control Act of December 22, 1944, approving the Pick-Sloan Plan. He stated: "My approval of this bill is given with the distinct understanding that it is not to be interpreted as jeopardizing in any way the creation of a Missouri Valley Authority, the establishment of which should receive the early consideration of the next Congress."[140] Roosevelt's MVA never materialized due to a lack of political support within Congress, though one last-ditch attempt was made in 1945.

The dedication ceremony for Glendo Dam and Powerplant featured, left to right, Dedication Queen, Jane Waters; Miss Cheyenne, Vedana Blomquist; Miss Casper, Marilyn Schipporeit; Miss Wyoming, Linda Phillips; and Mayor of Glendo, Roy Amick. Pick-Sloan Missouri Basin Program: Glendo Unit, Wyoming. Date: June 9, 1959. Photographer: Unknown.

Together, under the auspices of the PSMBP, Reclamation and the Corps constructed or investigated approximately 150 possible damsites. The benefits of these water development projects included navigation, flood control, irrigation, municipal and industrial water supplies, preservation and enhancement of fish and wildlife, recreation opportunities, power generation from plants with a total installed capacity of about 2.5 million kilowatts, and other uses. Reclamation integrated the power system of the Colorado-Big Thompson, Kendrick, North Platte, and Shoshone Projects with the PSMBP transmission system. Reclamation then collectively marketed the power produced from these projects. Through the sale of energy, the program returns revenues sufficient to cover the annual operating expenses for power facilities and a reserve for their future replacement, while at the same time repaying a portion of the construction costs of the facilities. Using the knowledge gained through the development and interconnection of the Pick-Sloan power system, Reclamation then turned its attention to the Upper Colorado River Basin.

THE UPPER COLORADO RIVER BASIN

In 1904, Reclamation began investigating the Flaming Gorge area in Wyoming to provide flood control storage in the Upper Colorado River Basin and to prevent flooding of the

Flaming Gorge Reservoir. Colorado River Storage Project: Flaming Gorge Unit, Wyoming, Utah. Date: Unknown. Photographer: Deborah Linke.

Lower Basin below Black Canyon, the present site of Hoover Dam. Reclamation continued investigating in more detail in 1914, but nothing came of these early investigations, and they were discontinued.

Reclamation began envisioning a need for the Colorado River Storage Project (CRSP) in 1922, as the basin states began negotiations of what eventually became the Colorado River Compact.[141] The allocation of 7.5 million acre-feet of water per year to each of the Upper and Lower Basins, as decreed in the Compact, hinged on the Upper Basin's

Upper Molina Powerplant. Collbran Project, Colorado. Date: October 15, 1998. Photographer: Bradford P. Osburn.

delivery of an average of seventy-five million acre-feet of water in any ten-year period, with additional water for use in Mexico. Committed by the Compact to providing a fixed amount of water to the Lower Basin on a regular basis, the dilemma for the Upper Basin became how to meet Compact obligations and retain enough water for its own development.

As with most western rivers, the flow of the Colorado River varies erratically from year to year, ranging from five to twenty-four million acre-feet per year at Lee Ferry, Arizona. In addition, high and low water years tend to follow one another, accentuating problems of river regulation and use. Prolonged dry periods pose problems for the Upper Basin, both in meeting their Lower Basin delivery requirements and utilizing their own allocations. To ensure availability of the Upper Basin's allocation in all years, large storage reservoirs needed to be built. Filling reservoirs during wet years would allow the Upper Basin to meet its delivery requirements to the Lower Basin in dry years, while still providing for Upper Basin water development.

A scarcely tapped potential of agricultural, industrial, natural resources, and recreational assets existed in the Upper Colorado River Basin. The basin contained tremendous quantities of uranium, coal, and other minerals. Realization of the potential economic growth and contribution to the national welfare depended upon the maximum effective use of limited water supplies. Reclamation designed the Colorado River Storage Project to conserve the very limited precipitation, which falls primarily in the form of snow in the high mountains, and use it for natural resources development and municipal, industrial, and agricultural growth in the Upper Colorado River Basin. The proposed project would also provide flood control, extensive recreation, and fish and wildlife preservation.
Before passage and implementation of CRSP, several obstacles remained. The most significant of them was the

During a trip from Casper, Wyoming, to the Grand Canyon, Bert Hanna of the *Denver Post,* Secretary of the Interior Stewart L. Udall, and Commissioner Floyd Dominy toured Flaming Gorge Dam and Dutch John, Utah, with Project Construction Engineer Jean R. Walton. Colorado River Storage Project: Flaming Gorge Unit, Wyoming, Utah. Date: April 28, 1961. Photographer: F.B. Slote.

further allocation of the waters of the Colorado River. In 1949, Congress ratified the Upper Colorado River Basin Compact, which divided the Upper Basin's allotment of 7.5 million acre-feet of water among the four Upper Basin States (Colorado, New Mexico, Utah, and Wyoming).[142] The Upper Colorado River Basin Compact paved the way for further development of the Colorado River.

Construction of Crystal Dam. Colorado River Storage Project: Wayne Aspinall Unit, Colorado. Date: November 20, 1974. Photographer: Unknown.

Reclamation worked in conjunction with the Upper Colorado River Commission and other Federal entities to prepare a formal report in 1950, describing plans for a series of dams and reservoirs to augment development in the Upper Basin. This plan included power development and other services. According to the report, revenues from hydropower developed at the CRSP mainstem units would be used to fund smaller irrigation projects known as participating projects. The report outlined plans for an initial group of participating projects, developing water for irrigation and other purposes, that would be financially linked with the developing Colorado River Storage Project mainstem units.

With passage of the Colorado River Storage Project Act of April 11, 1956, Congress authorized construction of four storage units in CRSP and eleven participating projects, and Reclamation undertook the development of the upper half of a large river basin. The four mainstem units of CRSP – Glen Canyon, Flaming Gorge, Navajo, and Wayne Aspinall[143] – stretched from the Green River in Wyoming down the Colorado River and its tributaries to Page, Arizona. All but Navajo contained hydroelectric facilities designed to repay construction costs of the initial facilities, as well as costs associated with construction of several participating projects. The participating projects were comprised of: the initial phase of the Central Utah Project (CUP), Eden,[144] Emery County, Florida, Hammond, La Barge, Lyman, additional works at Paonia, Pine River Extension, Seedskadee, Silt, and Smith Fork Projects. Reclamation later found the La Barge Project in Wyoming and the Pine River Extension Project in Colorado infeasible and deleted them from the project. Additional participating projects later added by Congress were Animas-La Plata (1968); Bostwick Park (1964); Dallas Creek (1968); Dolores (1968);

Glen Canyon Dam. Colorado River Storage Project: Glen Canyon Unit. Date: Unknown. Photographer: Unknown.

Fruitland Mesa (1964); Navajo Indian Irrigation Project (1962), a Bureau of Indian Affairs project constructed by Reclamation; San Juan-Chama (1962); San Miguel (1968); Savery-Pot Hook (1964); and West Divide (1968).

The Flaming Gorge Unit and the Wayne Aspinall Unit helped provide necessary long-term regulatory storage of the Colorado and Green Rivers. This storage assisted the states in the Upper Basin of the Colorado River in meeting their flow obligations at Lee Ferry, as required under the terms of the Colorado River Compact and the various other compacts governing the river. The units also permitted the Upper Basin States to use their apportioned share of the river. Powerplants located at the dams – Flaming Gorge, Blue Mesa, Crystal, and Morrow Point – produced a portion of the electrical energy needed in the Upper Basin. Revenues from power development on the project repaid hydroelectric costs. Any remaining funds were then applied to irrigation costs incurred by the participating projects that were beyond the repayment abilities of the project irrigators.[145]

The Navajo Unit helped further manage the water resources of the Upper Colorado River Basin. By providing irrigation waters for lands on the Navajo Indian Irrigation Project, the Navajo Unit helped to meet irrigation requirements for the Navajo Nation, while at the same time allowing the Secretary of the Interior to meet trust responsibilities to that tribe. In addition, water releases helped to maintain a continuous flow of water for power generation at Glen Canyon Dam.[146]

The Glen Canyon Unit provided long-term storage for the Upper Basin, allowing it to meet yearly flow requirements at Lee Ferry. In addition, Glen Canyon Powerplant numbered among Reclamation's five largest hydropower plants. The plant provided a significant source of revenue for development of the Upper Basin, as well as electrical energy to aid in that development.

THIRD POWERPLANT AT GRAND COULEE

The Colorado River was not the only western river that required long-term planning prior to its development. The Columbia River also required considerable planning before it could be fully developed.

Prompted by postwar power shortages in the Pacific Northwest, Reclamation began construction on Hungry Horse Dam in Montana in 1944. Realizing that the powerplant at Hungry Horse would not be sufficient to address the growing energy demand in the region, Reclamation began exploring other alternatives.

The largest obstacle to additional hydropower construction in the Pacific Northwest was the lack of year-round, predictable water flow of the Columbia River. To justify additional powerplants at Grand Coulee and other

Glen Canyon Dam under construction. Colorado River Storage Project: Glen Canyon Unit, Arizona, Utah. Date: April 9, 1963. Photographer: Unknown.

Generator inside Blue Mesa Powerplant. Colorado River Storage Project: Wayne Aspinall Unit. Date: July 2001. Photographer: T.R. Linenberger.

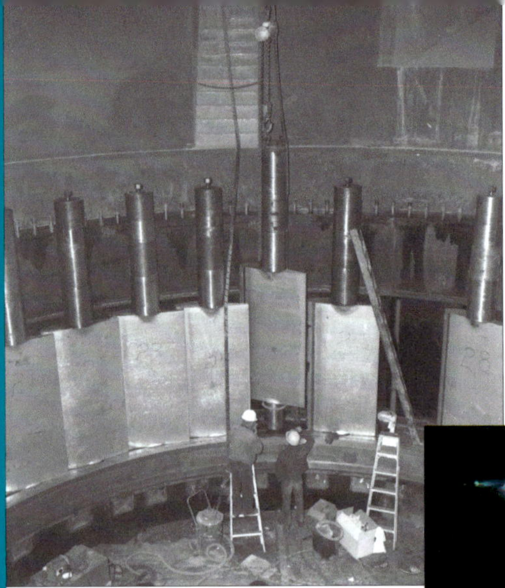

Installing the wicket gates on Unit No. 19. Columbia Basin Project: Third Powerplant, Washington. Date: November 19, 1974. Photographer: H.S. Holmes.

Third Powerplant at night. Columbia Basin Project, Washington. Date: Unknown. Photographer: Unknown.

Third Powerplant at night. Columbia Basin Project, Washington. Date: Unknown. Photographer: Unknown.

Unit 21 generator shaft and thrust bracket installed. Columbia Basin Project, Washington. Date: May 24, 1976. Photographer: Bob Isom.

Unit 22 turbine runner. First blade-to-band weld just completed. Columbia Basin Project: Third Powerplant, Washington. Date: May 15, 1976. Photographer: H.S. Holmes.

View showing construction progress of the superstructure. Columbia Basin Project: Third Powerplant, Washington. Date: June 13, 1972. Photographer: Bob Isom.

View of the partially constructed superstructure. Columbia Basin Project: Third Powerplant, Washington. Date: June 27, 1972. Photographer: H.S. Holmes.

A general view of the Third Powerplant and Forebay Dam as seen from the temporary access road to the right powerplant. Columbia Basin Project, Washington. Date: January 25, 1972. Photographer: Bob Isom.

Third Powerplant at Grand Coulee. Columbia Basin Project, Washington. Date: Unknown. Photographer: Unknown.

sites, Reclamation needed to provide additional storage capacity to balance the river's flow throughout the year.

However, the waters of Lake Roosevelt behind Grand Coulee Dam already backed nearly into Canada. Because of the Corps' dams downstream, any new structures had to be built upstream, putting them in Canadian territory.

While the idea of building dams and reservoirs on Canadian rivers seemed logical from an engineering standpoint, the politics were complex. The international nature of the project required that the United States government become involved. In a process that took almost two decades, the two governments negotiated the Columbia River Treaty.

In March of 1944 – even as construction began on Hungry Horse Dam – the United States government referred the issue of additional storage on the Columbia River to the International Joint Commission (IJC), a six-person body that gives final approval to any project or case involving the use, obstruction, or diversion of any boundary waters between the United States and Canada. The IJC, in turn, created the International Columbia River Engineering Board, comprised of two members from each country, which set up an engineering committee to collect data and analyze the situation. These extensive technical studies took roughly fifteen years to complete. From the onset of the formal planning process, the benefits of additional storage to the United States were evident. Among other things, the regulated flow of the Columbia would reduce flooding and allow for additional and consistent hydropower generation. In May of 1948, widespread damage due to flooding throughout the Columbia Basin proved the catalyst that prompted action by the various planning bodies. The flood also increased Canadian willingness to accommodate storage in return for financial compensation or electricity.

In a revised "308 Report" on the Columbia Basin, published in October of 1948, the Corps recommended construction of twelve to fifteen additional dams on the upper Columbia River and its tributaries, most of which were never built. One that did make it out of the planning stages, Libby Dam located on the Kootenay River near Libby, Montana, needed the approval of the Canadian government before construction could begin, because forty-two miles of the planned reservoir extended into Canada. The United States Congress granted approval for the project in the omnibus River and Harbors-Flood Control Authorization Bill of 1950. The United States then urged the IJC to approve the project. The IJC held hearings on the project, but, in 1953, Canada denied the federal license required for construction of such projects, and the United States withdrew its application.

At the same time that the Corps proposed Libby Dam, plans moved forward for additional structures upstream of Grand Coulee Dam, including several located entirely in

Looking down from the crest of Hungry Horse Dam on the powerplant and the river below. Hungry Horse Project, Montana. Date: August 13, 1975. Photographer: Glade Walker.

Toby Levy, an engineer at the Third Powerplant. Columbia Basin Project, Washington. Date: July 28, 1970. Photographer: Unknown.

Canada. A 1949 preliminary report by the Columbia River Engineering Board recommended construction of a dam at Arrow Lakes, just north of the international border near Castlegar, British Columbia. The report stated that the single dam would not end downstream flooding. When federal plans for the structure stalled, several private entities, including the Kaiser Aluminum and Chemical Corporation, attempted to construct the dam. These private efforts also met with opposition from the Canadian government in Ottawa, and the upper Columbia remained largely undeveloped.

Flooding and regulation were not the only issues at stake on the Columbia River. Even as the final generator at the Grand Coulee Powerplant went into operation in 1951, Reclamation and Washington promoters advocated additional power production on the Columbia.[147]

A large portion of the problem with development of the Columbia stemmed from internal conflict within Canada. British Columbia supported the United States proposals because it thought the construction would aid in growth of the province. Ottawa, home of the national government, was more reluctant to sanction development. Continued opposition from Ottawa prompted the province to consider local development and to court private United States developers, including Kaiser, which led to the Arrow Lakes proposal.

Changing political winds in both countries played a significant role in the progress of development of an international treaty. In the United States, the pro-construction Truman Administration gave way in 1953 to the conservative Eisenhower Administration and the "no new starts" policy. In Canada, the Liberal leadership of Louis St. Laurent yielded to the Progressive Conservative Prime Minister John G. Diefenbaker in 1957.

Even as negotiations stalled in 1954, the United States resubmitted a proposal to construct Libby Dam. The proposal suggested a new location several miles upstream that would inundate far less Canadian land. Calling for additional benefits, the Canadian government suggested a diversion of roughly sixty-seven percent of the Columbia's flow at Revelstoke Lake and into the Fraser River to the north.[148] The diversion would significantly decrease the amount of water flowing downstream into the United States. The Canadian Parliament proved its seriousness in this matter by appropriating money to study the feasibility of diverting both the Columbia and the Kootenay.[149]

Though probably designed to unify the various Canadian entities, the diversion plan in fact only served to further alienate the British Columbia government. Further attempts to privately construct dams along the Columbia met with Parliamentary resistance, forcing British Columbia to consider local development of the Peace River to satisfy provincial energy needs. This move effectively sidestepped national authority.[150]

In the midst of the complicated negotiations between the United States and Canada, another factor came into play, further underscoring the importance of the negotiations. In 1955, the then Soviet Union built the Krasnoyarsk Powerplant on the Yenisei River in Siberia. In a Cold War indignity for the United States, the combined output of the plant's generators topped that of Grand Coulee, previously the world's largest hydropower plant. In a further affront, the Russian generators had a larger capacity than those operating at Coulee. The loss of all its first place hydropower claims, combined with the embarrassment of Sputnik in 1957, may have provided added impetus for construction of a Third Powerplant at Grand Coulee.

For the next several years, each interested entity continued to press its solution for development of the Columbia River. Finally in 1958, President Eisenhower and Prime Minister John Diefenbaker met and agreed to do something about the river. Bypassing the IJC, the two leaders scheduled a series of meetings beginning in February of 1960. The Americans still advocated construction of Libby Dam, while the Canadians preferred storage on the Kootenay River. Ultimately, the compromise agreement called for construction of Duncan Lake, High Arrow, and Mica Dams in Canada and Libby Dam in the United States.[151] The combined storage equaled twenty million acre-feet, more than enough to contain floods rivaling the 1948 disaster. The agreement further stated that the United States would pay one-half the cost allocated to flood control benefits, roughly sixty-five million dollars, and Canada would not divert Columbia River water for at least sixty years.

President Eisenhower officially announced the agreement in October of 1960. On January 17, 1961, Eisenhower performed one of his last official acts as President when he and Diefenbaker signed the Columbia River Treaty in Washington, D.C. The U.S. Congress easily ratified the treaty. The Canadian government, however, did not prove quite so accommodating.

The ongoing conflict between British Columbia and Ottawa was the treaty's undoing in Canada. British Columbia thought that Ottawa overlooked the province's goals. The provincial government rejected the use of American funds to construct the storage reservoirs and expressed concern over recovery of the Canadian share of the jointly produced power. It took another two years for the two sides to reach a meeting of the minds, partly attributable to Lester Pearson being elected Canada's new Prime Minister.

Diplomats reopened negotiations in December of 1963 and quickly reached agreement. The Canadian Parliament ratified the treaty on June 10, 1964, and the U.S. Congress quickly followed suit. President Lyndon Johnson and Prime Minister Pearson signed the new Columbia River Treaty and Protocol on September 16, 1964, at

President Harry S. Truman throws a switch to signal the start of the first power production at the Hungry Horse Powerplant, during dedication ceremonies held in Flathead County High School gymnasium at Kalispell, Montana. Hungry Horse Project, Montana. Date: October 1952. Photographer: A.E. McCloud.

Vanport, Oregon, a housing project built for defense workers during World War II, after it had been ravaged by flood waters. Date: June 18, 1948. Photographer: Stanley Rasmussen.

Each blue circle marks a hydropower generating unit on the floor of Grand Coulee's Third Powerplant. Columbia Basin Project, Washington. Date: Unknown. Photographer: Unknown.

the Peace Arch in Blaine, Washington, almost twenty years after negotiations between the two governments began.

The signing of the treaty and the resulting construction of storage reservoirs to balance the flow of the river finally made a Third Powerplant at Grand Coulee feasible. The additional storage provided by the new reservoirs in Canada allowed for the further regulation of the Columbia to ensure a consistent flow of water through the turbines in the Third Powerplant.[152] At the same time, by guaranteeing the necessary water supply, it paved the way for a Pacific Northwest-Pacific Southwest Intertie system stretching from the Pacific Northwest to the Southwestern United States. Congress authorized construction of the Intertie in August of 1964, just a month prior to the official signing of the treaty.

With the treaty negotiated and the issues between Canada and the United States resolved, the United States turned its attention toward authorization and construction of a Third Powerplant at Grand Coulee. Secretary of the Interior Stewart Udall approved Reclamation's tentative plans for the project in February of 1965. The Senate followed suit and authorized the project four months later, although disagreement within the U.S. House of Representatives delayed authorization of the project for another year. Finally, on June 14, 1966, President Lyndon Johnson signed a bill authorizing construction of the Third Powerplant in a White House Rose Garden ceremony.

Construction of the Third Powerplant began in 1967, after Secretary Udall announced the new powerhouse would use 600,000 kW generators instead of 300,000 kW ones.[153] This change would make Grand Coulee the largest powerplant in the world, as well as home to the largest generators in the world. The specifications for the first three units in the Third Powerplant required generators capable of producing 615,385 kVA (600,000 kilowatts). Prior to issuing the specifications for the last three units, Reclamation engineers realized that by using stator windings that were water cooled (similar to those used in steam turbines) it would be possible to increase the capacity even further. Therefore the specifications for the last three units required a continuous overload capacity of 825,640 kVA (805,000 kilowatts). These specifications automatically meant that the units would be breaking technological ground, and Reclamation employees would have to solve any problems that developed along the way.

Udall's announcement caused an unexpected stir among the international community. Not long after the announcement, the Soviet Embassy phoned the Department of the Interior to inquire if the powerhouse contracts would be open to international bidding, and pointed out that they had already constructed 500,000 kW generators. If the Soviets won the bid, they could claim at least a portion of the world's largest powerplant, a potentially embarrassing proposition for the United States in the midst of the Cold War.

Somewhat surprisingly, opinion was divided on the idea of Soviet involvement in the project. President Johnson and the State Department argued that allowing the Soviets to bid on at least half of the units would be a potential first step in easing Cold War tensions. Stewart Udall and Congress believed the United States should construct the first three – but preferably all – units.[154] The Department of the Interior effectively ended all discussion on the matter by announcing only domestic bids would be accepted. Interior further stipulated that the units must contain at least fifty-one percent domestic parts. Delays in construction rendered the point moot, as Cold War tensions subsided, and foreign involvement became less of an issue.

Enlarging the generators at the Third Powerplant was not as easy as just announcing the change in the contract specifications. To make the generators effective within the existing system, Reclamation had to modify existing technologies. Previously, Reclamation and the industry used two or three cycle power circuit breakers; however, the new larger generators at Coulee necessitated faster one-cycle circuit breakers.

Reclamation's laboratory staff field tested the new circuit breakers, thus identifying changes that needed to be made by the manufacturer to meet power system stability requirements. The one-cycle breakers separated the generators from the power system much more quickly, shortening the duration in which the fault current from the generator fed into the power system. This had the potential to reduce the damage caused during an electrical fault and also decreased the likelihood that the power system would become unstable. The early designs for the first one-cycle breakers later proved faulty, and a second series of Reclamation-led tests to identify the weakness was performed. This time, manufacturer modifications reduced the breaker capabilities.
In addition, technology had moved forward, and the original fastest, largest breakers were replaced with newer technology.

Reclamation awarded the first contract for construction of the Third Powerplant in 1970, having previously begun earth removal and other preliminary jobs at the dam. Part of the construction included removing a section of the existing dam, a significant undertaking. In the midst of construction activities, President Nixon announced budget cuts aimed at reducing government construction activities by seventy-five percent. The urgent need to meet Pacific Northwest power needs and fulfill Canadian treaty agreements allowed the government to exempt Grand Coulee from the budget cuts, and construction remained relatively on schedule. Though several delays occurred due to labor disputes and contractor difficulties, the plant was finished only a year behind schedule. The first of the six new generators went online in 1975, the last in 1980.

Reclamation Electric Power Branch personnel conducting a line-to-line fault test on the 345 kV Glen Canyon-Pinnacle Peak Transmission line. This verified the effectiveness of protective devices in the system. Pacific Northwest-Pacific Southwest Intertie, Arizona. Date: May 1970. Photographer: Unknown.

The switching station for the Los Angeles Bureau of Power and Light transmission line, near Hoover Dam. Boulder Canyon Project, Arizona, Nevada. Date: May 20, 1936. Photographer: G.L. Ullom.

Grand Coulee Dam. Columbia Basin Project, Washington. Date: Unknown. Photographer: Deborah Linke.

The United States again claimed a Cold War victory by surpassing the amount of power generated at the Soviet's Krasnoyarsk Station. The Third Powerplant represents one of the best examples of Reclamation's vital role in power generation. Grand Coulee's power capability allowed it to serve not only the Pacific Northwest, but potentially the entire trans-Rocky Mountain West through a Pacific Northwest-Pacific Southwest high-voltage transmission system.[155]

THE PACIFIC NORTHWEST-PACIFIC SOUTHWEST INTERTIE

Long before completion of the Third Powerplant at Grand Coulee, Reclamation recognized a need for the regional exchange of both public and private power. After years of debate between public and private interests, Reclamation's effort was one of the first attempts to provide direction for the large-scale integration of both public and private utility systems. The proposed connection of the Pacific Northwest and Southwest would have alleviated existing seasonal excess electric power capacity by transferring energy between the two areas. When the summer heat in the Southwest increased air conditioning use, surplus electric power capacity in the Northwest could be called on to meet the higher demand in the south. The reverse could occur in the winter months when energy use in the Northwest increased due to heating needs and decreased in the Southwest as the weather cooled.

The process leading to authorization of the Pacific Northwest-Pacific Southwest Intertie and the start of construction activities took nearly thirty years. Much of the early work on the project occurred during the struggle between the public and private power interests in the 1930s, giving the Intertie a dramatic history, as pointed out in Reclamation's first annual project history for the Intertie.

> This capsule history of the struggle for the Intertie makes no attempt to record the drama, the elements of controversy which on many occasions nearly scuttled the program, the give-and-take negotiations, the long hours of work by many people, or the leadership of the President [John F. Kennedy], the Secretary [of the Interior Stewart Udall], and many members of Congress which led to the eventual agreement.[156]

Initial investigations for an interconnected system began as early as 1935, when the Pacific Northwest Regional Planning Commission issued a report entitled, "The Columbia Basin," which envisioned the eventual interconnection of power resources in the far West. In 1949, Reclamation conducted the first detailed investigation of a possible interconnection between the

Bonneville system and the Central Valley Project of California. This investigation found that a 230,000 volt, 217 mile connection to close the gap between Roseburg, Oregon, and Shasta Dam, was economically feasible and desirable. A Federal Power Commission study in 1953 reaffirmed the economic advantages of a strong interconnection between the two regions, comprised of either one or two 230,000 volt connections. Soon after the Federal Power Commission study confirmed the value of a connected system, the Bonneville Yamsay-Klamath Falls line was sold to the California-Oregon Power Company (Copco), preventing the Bonneville system from reaching the California border as originally planned and forestalling any attempts to connect the Central Valley Project and the Bonneville system. Meanwhile, BPA's surplus hydropower capacity continued to go unused, causing concern over the organization's financial security.

In April of 1959, Secretary of the Interior Fred Seaton directed BPA and Reclamation to study the California Intertie to dispose of surplus power capacity. Meanwhile, in 1959 and 1960, Pacific Gas and Electric Company proposed a 230,000 volt connection. This proposal was deferred by the Senate Interior Committee, pending passage of legislation assuring each region that their power consumers would have first priority for the Federal hydroelectric power generated in their respective areas.

The present Intertie system resulted from President Kennedy's message to Congress on natural resources on February 23, 1961:

> Finally, I have directed the Secretary of the Interior [Stewart Udall] to develop plans for the early interconnection of areas served by that Department's marketing agencies with adequate common carrier transmission lines; to plan for further national cooperative pooling of electric power, both public and private; and to enlarge such pooling as now exists.[157]

The Kennedy Administration further encouraged power pools.[158] On March 10, 1961, Secretary Udall appointed a special task force, headed by BPA's Administrator Charles Luce, to report on the possibilities of connecting BPA and Reclamation's Central Valley Project. In December of that same year, the task force reported that "an extra high voltage interconnection between the two regions should be constructed at the earliest practicable time."[159]

In January of 1962, the President's budget message included a request for funds for the design and construction of an extra-high voltage connection between the Central Valley Project and the Bonneville Power Administration system, prompting Congress to provide $300,000 for continuing studies. Concurrently, regional protective

Jamie Damon, 9, of Pierre, South Dakota, greets President John F. Kennedy at the powerplant at Oahe Dam. To the left and rear of President Kennedy is Lt. General Walter K. Wilson, Jr., Chief of Army Engineers. George McGovern, former Director of Food for Peace, looks on from the right of the photo. Date: August 17, 1962. Photographer: T.R. Broderick.

legislation was introduced to Congress. The Senate passed the legislation, but it stalled in the House Interior Committee. The bill was reintroduced in 1963, and funds were requested to begin construction on the Intertie lines. In November, Congress appropriated seven million dollars for construction, pending approval of the regional protective legislation. By this time, the bill had passed both Houses but in different forms, which needed to be reconciled in Conference Committee.[160]

Helicopters are used to inspect transmission towers and transmission lines near Hoover Dam. Boulder Canyon Project, Arizona, Nevada. Date: August 12, 1947. Photographer: William S. Russell.

On June 24, 1964, nearly thirty years after the original idea for an interconnected system was conceived, Secretary Udall again submitted the proposal for the present Pacific Northwest-Pacific Southwest Intertie to Congress. Congress approved the final version of the plan in August and appropriated funds to begin construction on the Federal portions. The same day, the Senate-House Conference Committee reached an agreement on the regional protection legislation. Passage of the bill followed a week later, paving the way for the final authorization of the Pacific Northwest-Pacific Southwest Intertie on August 31, 1964. Construction activities began soon after.[161]

While Congress was debating the merits of an Intertie, the rapidly growing electric power system in the West began experiencing power oscillations of unknown origin. Whenever a large amount of energy was being transferred from the Pacific Northwest to the Southwest, energy would begin to swing back and forth across the transmission lines until the transfer level was reduced. With an abundance of cheap energy in the Northwest and growing loads in the Southwest, this problem had to be solved. Reclamation research engineers, in cooperation with engineers from utilities throughout the West, began to study the phenomenon. After several months of study and field tests, the problem was traced to interaction between generator voltage regulators and the long transmission lines of the West.

As it turned out, Canadian engineers were also studying the same types of problems and arrived at the solution at the same time as Reclamation engineers were producing the first "Supplementary Excitation Controller." Later renamed the Power System Stabilizer, this shoebox-sized device filled with electronic circuits monitored the frequency of the power system and sent a damping signal to the voltage regulator whenever power and frequency oscillations began. Installation of these devices at several key locations around the power system enabled much more of the excess energy production capacity from the Northwest to be transferred to the Southwest. These devices are now required on large generators throughout the world where long transmission lines are encountered.

The Pacific Northwest-Pacific Southwest Intertie system was originally conceived as a multiple transmission line system, including several substations. As originally planned, the Intertie was to be comprised of several shorter alternating current transmission lines, as well as two longer direct current transmission lines. The shorter lines would have run from Oregon to California, and Arizona, while the two longer lines would have originated in Oregon with one dc line to Los Angeles, California, and the other to a point near Phoenix, Arizona. Due to a reduction in funding and regional political pressures, Reclamation built the ac lines but not the dc ones. The Los Angeles Department of Water and Power (LADWP) and BPA later built one of the planned dc lines from Celio, Oregon, to Sylmer, California; however, the entire Intertie system was never completed as originally designed. Reclamation transferred the Intertie system and the rest of its transmission capabilities to the Western Area Power Administration (WAPA or Western) in compliance with the 1977 Department of Energy Organization Act.

The Intertie connected private, state, and Federal power systems. These systems included the largest Federal hydropower system in America that sold energy produced by the Corps and Reclamation (BPA); the largest municipal system (the Los Angeles Department of Water and Power); and the largest group of private systems in the West (the California Power Pool, comprised of the Pacific Gas and Electric Company, the Southern California Edison Company, and the San Diego Gas and Electric Company).[162] Reclamation hydroelectric powerplants in the system generated energy that was subsequently distributed through the Intertie and the western area power grid to eleven western states: Arizona, California, Colorado, Idaho, Montana, Nevada, New Mexico, Oregon, Utah, Washington, and Wyoming.[163] Currently, the Intertie is jointly controlled; BPA controls the portion of the project located in Oregon, while Western controls the rest of the system.[164]

Electrical energy consumers receive the greatest benefits from the Intertie system. The construction of the additional power lines, the interconnection of the different hydroelectric systems, and the interconnection of steam generating systems has increased the availability, dependability, and stability of power as well as the means by which it is received. The Intertie can deliver excess capacity to customers who otherwise may have experienced frequent disruptions of power and helped alleviate energy shortages in southern California in 2001.

The Intertie provides an example of the successful cooperation between public and private utilities, despite their long history of friction. The original plan called for the seasonal exchange of energy between the Pacific Northwest and the Pacific Southwest, as well as public and private utilities. Though its original plan changed as construction progressed, the Intertie remains an integral part of the transmission network in the West. The Intertie connects many of the major utilities, providing for sharing power when

Spillway section of Shasta Dam with gates in operation. Central Valley Project: Shasta/Trinity Division, California. Date: July 18, 1951. Photographer: T.B. Gibson.

necessary and assisting during emergencies. In addition, the Intertie comprises a component of the network of transmission lines owned and operated by the Western Area Power Administration since 1977.

DEVELOPMENT OF DIRECT CURRENT TRANSMISSION LINES

Transmission towers stretching across the Columbia Basin. Columbia Basin Project, Washington. Date: April 21, 1979. Photographer: Glade Walker.

Before actual construction on the Intertie could begin, Reclamation had to research the transmission of extra high voltage electric power. To design and plan the Intertie, Reclamation created a miniature model of various aspects of power systems. The miniature system allowed engineers to determine the functional characteristics of the alternating current system before construction even began. In conjunction with the miniature ac system, Reclamation built a dc model of the circuitry of the Intertie to test the compatibility of the two systems.

The transmission distances encountered with the Intertie were two to three times greater than those previously experienced. Because of the greater distances, existing technology had to be expanded to accommodate the new dc lines and ensure compatibility with both old and new ac lines.

Direct current technology had been around for many years prior to plans to develop the Intertie. In 1905, Dr. C.P. Steinmetz, then of the General Electric Company, developed mercury rectifiers (devices that converted ac to dc) to supply street lighting in Schenectady, New York. After several years of refinements, "that company's 17-mile long direct current transmission line from Mechanicville to Schenectady, New York, operated from 1936 to 1945."[165] Though dc was once the established means of delivering and using electric power, technical difficulties limited its use. For this reason, the commercial and industrial use of electricity began switching over to ac at the end of the nineteenth century. Interestingly enough, "among the last holdovers were low-voltage dc distribution systems serving the trolley cars and buses then in operation."[166]

While the use of dc in the Intertie system was limited, Reclamation undertook a number of technical studies to increase the available knowledge of dc, to learn the advantages and disadvantages of using dc for the Intertie, and to obtain firsthand experience with its use, particularly over long distances and at extra high voltage.[167] The proposed Intertie, based on Reclamation research, combined the best features of the ac and dc systems.

In the United States the major advantage to using a dc transmission line for long distance transmission was economic. At the same time, dc provided for maximum energy transfer through a given transmission line right-of-way.

To use direct current technology, Reclamation engineers had to design different towers for a dc line. Originally, tower designs were based on the use of two single wire conductors, with one overhead ground wire for lightning protection, and an anticipated spacing of 1,100 feet between towers. Design requirements were revised when engineers found that greater current-carrying capacity could be provided with a "bundled" conductor. Using the bundled conductor, towers could be placed fifty feet farther apart, conserving construction time and money. These changes, in turn, called for a stronger tower to support the added weight.[168]

The significant technological aspect of the Intertie was the proposed use of high-voltage direct current. From an engineering standpoint, the proposed use of dc is an important milestone in the field of American electrical development. New developments in dc broke the distance barrier in transmission, making it possible to efficiently move large quantities of energy great distances at low cost, which in turn made it feasible to generate power at isolated locations.[169]

The use of dc was new to Reclamation engineers investigating its use as part of the Intertie. Switzerland, Germany, Sweden, France, England, the U.S.S.R., New Zealand, Italy, Japan, Denmark, and Canada had all made advances in dc electrical systems and transmission lines that could then be used by Reclamation engineers working on the Intertie.[170]

Development of the North American Electric Reliability Council

As Reclamation investigated and started building the Intertie, a reliability crisis began to develop throughout the electric utility industry. Reliability and regulation met head-on in the great northeastern blackout of 1965, prompting a voluntary, industry-driven shift in national policy for electric utilities.[171]

In November of 1965, a single electrical surge on the power grid brought down the eastern electric system, including the entire Northeastern United States and Ontario, Canada. This failure is most commonly known as the New York City blackout, but it extended well beyond that city, lasted for roughly twelve hours, and left approximately thirty million people in the dark. The blackout emphasized America's dependence on electricity for daily life and the unreliability of its electrical systems. The electric utility industry responded to the crisis by forming the North American Electric Reliability Council (NERC) and regional reliability councils. NERC was created in 1968; several of the regional reliability councils were formed prior to the establishment of NERC.

A strictly voluntary, not-for-profit organization, NERC was formed to assure the reliability of the North

**Navajo Generating Station.
Date: February 11, 1989.
Photographer: Gene Hertzog.**

American electrical system. Ten regional councils[172] constituted its membership. Essentially, NERC strove to prevent future large-scale blackouts by developing policies, standards, principles, and guides; monitoring the industry for compliance; and assessing future reliability of the entire system. To accomplish this goal, NERC worked with various segments of the electric utility industry, as well as its customers.

Reclamation and the power marketing administrations belong to the Western Systems Coordinating Council (WSCC).[173] The WSCC encompasses the entire western portion of the United States (the dividing line runs through the eastern edge of New Mexico, along the border of Colorado, the western edge of South Dakota, and the eastern edge of Montana) and the Canadian provinces of British Columbia and Alberta. Founded in 1967, WSCC addresses many of the same issues as NERC in a more regional manner.

New Waddell Pumping-Generating Plant. Central Arizona Project, Arizona. Date: July 19, 1993. Photographer: Joe Madrigal, Jr.

As the end of the twentieth century neared, NERC took steps to ensure its continued presence in, and importance to, the utility industry. To these ends, NERC and WSCC began shifting from a completely voluntary organization to a mandatory one through the added support of both the United States and Canadian governments.[174] Both WSCC and NERC are currently being reorganized with an independent board of trustees.

DEPARTMENT OF ENERGY

The Department of the Interior established the Southwestern Power Administration (Southwestern or SWPA) in 1943 in preparation for the transition to a peacetime economy, while continuing to meet wartime needs.[175] In 1950, the Secretary of the Interior created the Southeastern Power Administration (Southeastern or SEPA) to transmit and dispose of surplus power generated at Corps projects in the states of West Virginia, Virginia, North Carolina, South Carolina, Georgia, Florida, Alabama, Mississippi, Tennessee, and Kentucky. The Secretary of the Interior created the Alaska Power Administration (APA) in 1967 to operate and market power produced at Federal facilities in Alaska.[176]

By the mid-1970s, Reclamation had constructed an intricate power network, comprised of interconnected transmission lines, substations, switchyards, and generating facilities that traversed the entire western United States. To emulate this network throughout the remainder of the United States, President Jimmy Carter proposed legislation in 1977, which included creation of the Department of Energy (DOE), to further regulate the electric industry. Carter's proposal removed Reclamation's marketing and transmission responsibilities and placed all of the established Federal marketing bureaus – APA, BPA, Southeastern, and Southwestern – under the new department's jurisdiction. Congress easily passed the proposed legislation – the 1977

Department of Energy Organization Act – during the summer of 1977; President Carter signed it into law August 4, 1977.

In 1977, through the newly established legislation, President Carter established Western Area Power Administration to serve the remaining western states – Arizona, California, Colorado, Kansas,[177] Montana, Nebraska, Nevada, New Mexico, North Dakota, South Dakota, Utah, Wyoming, and part of Texas. The marketing bureaus and DOE were organized to separate the Federal power industry into two separate divisions – one responsible for the generation of power at hydroelectric facilities and one responsible for the marketing and transmission of Federal power to local utility companies.

The Corps and Reclamation retained responsibility for power generation at their individual facilities. The new marketing administrations within DOE took possession of most of the Federal transmission lines and other distribution facilities, assumed responsibility for their operation and maintenance, and accepted marketing responsibilities for the energy generated at Federal installations, surplus to Federal project needs, throughout the United States.

Western operates in the fifteen central and western states, with the exception of the areas where BPA operates, to coincide with Reclamation's own operating area; Western also operates in parts of Minnesota and Iowa.[178] Reclamation provided Western with a large percentage of its original staff, and the two bureaus have enjoyed a successful working relationship for more than twenty years. Western markets and transmits hydroelectric energy generated at fifty-five[179] powerplants operated by Reclamation, the Corps, and the International Boundary and Water Commission.[180] In addition, Western markets the United States' entitlement of the power generated at the non-federally owned, coal-fired Navajo Generating Station near Page, Arizona.[181] Western markets wholesale Federal energy to various utilities and organizations in the West; however, the preference customers who were eligible to receive Reclamation power did not change when Western took over Federal energy marketing. Western establishes power and transmission rates based on its own, Reclamation's, and the Corps' operation and maintenance costs, and fees to recover U.S. Treasury advances for project construction. Western returns the funds it collects to the U.S. Treasury to ensure Reclamation's repayment of the investment in Federal projects.

Creation of Western in 1977 largely coincided with the scaling down of Reclamation's large-scale construction activities and a completion, of sorts, of the power pools whose origins dated to the early days of Reclamation. Congress approved the last major multipurpose Reclamation projects in 1968, with the Colorado River Basin Project Act. This act authorized construction of several projects, including the Central Arizona Project in the Lower Colorado River Basin, which was declared substantially complete in 1993.

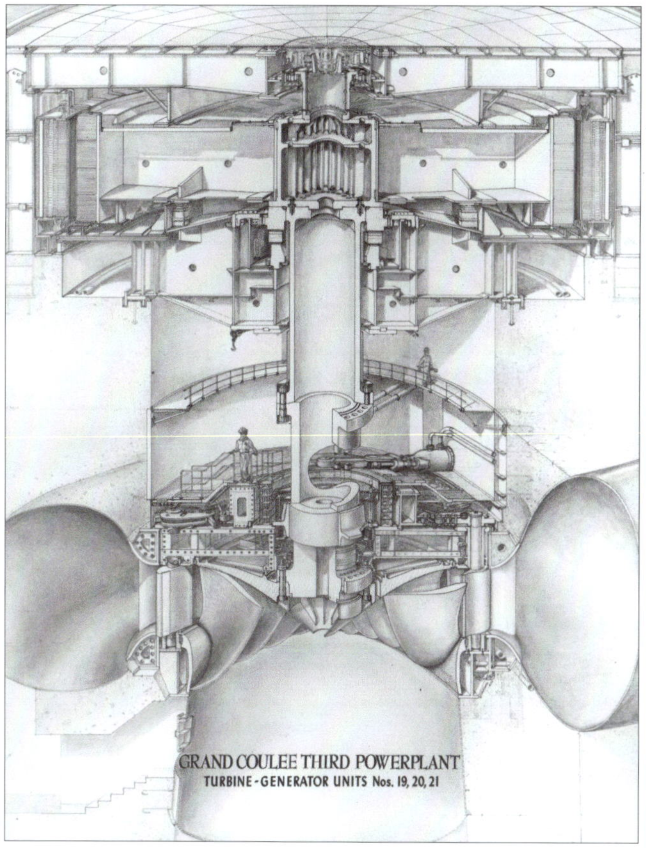

Cross section, Grand Coulee Third Powerplant - Turbine - Generator Unit Nos. 19, 20, 21. Date: April 1994. Artist: Jim Vitaliano.

Footnotes for Chapter 8: Reclamation and the Integration of Western Power Systems

[134] Corps, *History of the U.S. Army Corps of Engineers*, 2-15.

[135] Springer, 29-41; Gus Norwood, *Columbia River Power for the People: A History of Policies of the Bonneville Power Administration* (Portland, Oregon: Department of Energy, Bonneville Power Administration), 54-63; for an excellent and detailed discussion of this legislative process, see Phillip Funigiello, *Toward a National Power Policy: The New Deal and the Electric Utility Industry, 1933-1941* (Pittsburgh: University of Pittsburgh Press, 1973), 174-196.

[136] Originally called the Missouri River Basin Project, the project name was changed to the Pick-Sloan Missouri Basin Program in 1970, "in honor of two great men who contributed so much to the development of the water resources of the Nation." For the sake of consistency, the project is referred to by its current name, Pick-Sloan Missouri Basin Program, in this document.

[137] Golźe, 200-201; John E. Thorson, *River of Promise, River of Peril: The Politics of Managing the Missouri River* (Lawrence, Kansas: University Press of Kansas, 1994), 8, 14, 18; Michael L. Lawson, *Dammed Indians: The Pick-Sloan Plan and the Missouri River Sioux, 1944-1980* (Norman, Oklahoma: University of Oklahoma Press, 1982), 5.

[138] U.S. Army Corps of Engineers, *Big Dam Era: A Legislative and Institutional History of the Pick-Sloan Missouri Basin Program*, by John R. Ferrell (Omaha, Nebraska: U.S. Army Corps of Engineers, Missouri River Division, 1993), 3.

[139] Lawson, 19.

[140] Quoted in William E. Warne, *The Bureau of Reclamation* (Praeger Publishers, Inc., 1973; reprint, Boulder: Westview Press, 1985), 164.

[141] As discussed previously, the Colorado River Compact divides the waters of the Colorado River and its tributaries between the seven Colorado River Basin States.

[142] The Upper Colorado River Compact allocated the waters among the Upper Basin States in the following way: 51.75 percent to Colorado, 11.25 percent to New Mexico, 23 percent to Utah, and 14 percent to Wyoming. An additional 50,000 acre-feet was allocated to Arizona to compensate for existing use, as well as the minimal amount of water Arizona contributed to the Colorado River above the Lee Ferry dividing point. This translates to roughly 3,855,375 acre-feet for Colorado, 838,125 acre-feet for New Mexico, 1,713,500 acre-feet for Utah, and 1,043,000 acre-feet for Wyoming.

[143] Reclamation renamed the Curecanti Unit the Wayne Aspinall Unit on October 3, 1980.

[144] Even though construction on the project was well underway, the Eden Project was reauthorized in 1949, so it could be included in CRSP as a participating project.

[145] Denver, Colorado, National Archives and Records Administration: Rocky Mountain Region, Records of the Bureau of Reclamation, Record Group 115, "Annual Project History, Flaming Gorge Unit, Colorado River Storage Project, Utah-Wyoming," Vol II, 1957, 10; "Annual Project History, Flaming Gorge Unit, Colorado River Storage Project, Utah-Wyoming," Vol IV, 1959, 13-14; United States Department of the Interior, Water and Power Resources Service, *Project Data* (Denver: Government Printing Office, 1981), 358-360.

All project histories can be found in Record Group 115 at the National Archives and Records Administration – Denver, so hereafter will be noted by project.

146 Although unusual because it does not generate hydropower revenues to assist in Upper Basin development, the Navajo Unit provides much-needed water resources to the region.

147 Congress appropriated funds to Reclamation in 1952 to study the feasibility of an additional powerhouse at Grand Coulee.

148 Under then current international treaty provisions, Canada claimed it could do as it wished with the waters of the Columbia within its territory.

149 In his book on Grand Coulee, historian Paul Pitzer speculates that the suggestion to divert the Columbia was nothing more than a bluff on the part of the Canadian government, designed to emphasize its point. Whether serious or not, the Canadian government signaled its displeasure with the IJC negotiations and its desire for added benefits in return for providing the storage necessary to run American hydropower plants downstream. Pitzer, *Grand Coulee*, 336.

150 In 1955, the Canadian government passed the International River Improvements Act prohibiting construction of dams in Canada without federal approval. This move effectively negated any cooperative agreements between British Columbia and United States interests and further widened the gulf between the provincial and national governments.

151 B.C. Hydro completed the 1.4 million acre-foot Duncan Project on July 31, 1967, well ahead of its promised April 1, 1968, completion date. Keenleyside Dam, with 7.1 million acre-feet in Arrow Lakes, went into service October 10, 1968, also ahead of its April 1, 1969, promised completion date. Mica Dam, primarily designed to store water to generate power under the terms of the treaty, was completed three days prior to its promised completion date of April 1, 1973. The United States dedicated Libby Dam on August 24, 1975.

152 The United States (the Corps, Reclamation, and BPA) and Canada jointly manage and operate the flow of the Columbia River, under the Coordinated Columbia River System, for the benefit of the entire system. The Coordinated Columbia River System refers to projects operated under three separate but related arrangements: the Pacific Northwest Coordination Agreement, the Columbia River Treaty, and Federal flood control statutes. The Pacific Northwest Coordination Agreement was inspired by the Columbia River Treaty and specifically details the operation of both the American and Canadian Federal projects and generating utilities in the Pacific Northwest.

Under the Coordinated Columbia River System, the entities responsible for operation of dams along the river establish limits for each project as a group. These limits take into account the multipurpose needs of the river and its projects, as well as the Federal and international laws governing the use of the river, and allow for appropriate management of the dams on the river. In addition, the managers of the river attempt to provide consistent river flow for environmental needs, as well as maximum hydropower production.

153 Prior to the start of construction in 1967, Reclamation revised the construction plan to include three 600,000 kW generators and three 700,000 kW generators. Plans called for an additional six units that were never built or installed. The six constructed units added 3.9 million kW of capacity to the plant, bringing the nameplate capacity up to 6,163,000 kW. In 1964, Reclamation began rewinding the generators in the original two powerhouses. After completing all the rewinds in 1997, including the units in the Third Powerplant, plant nameplate capacity at Grand Coulee increased to 6,494,000 kW.

[154] In an oral history interview, Ted Mermel stated the following about the potential for foreign construction of the generators:

> Mermel: So at that time, Udall, in his bravado, said, "Well, we'll ask the Russians to furnish them." Then, that got into the newspapers, and it was in the papers, the Washington papers, that the Russians may bid on the Grand Coulee units. Naturally, the industry didn't like that and, at that time, mind you, Russia was not our ally at all. So this was rather offensive . . . This must have been about '66, '67, something in that era. So Dominy was continuing to press the Chief Engineer, saying, "Why don't you make some studies? Here Hitachi says they'll go for it . . ."

Oral history interview, Brit Allan Storey of Ted Mermel, January 26, 1995, Washington, D.C.

[155] Pitzer, *Grand Coulee*, 333-351; Norwood, 227-236, 247-251; Wm. Joe Simonds, *The Columbia Basin Project* (Denver: Bureau of Reclamation History Program), 1998.

[156] "Annual Project History, Region 2, Pacific Northwest-Pacific Southwest Intertie, Nevada-California," Vol I (1964), 11.

[157] Floyd E. Dominy, "A New Power Giant Materializes on the West Coast," *Reclamation Era* 51 (August 1965): 6.

[158] These pools linked transmission lines across regions and were encouraged by the formation of utility conglomerates that built nuclear plants. Only four existed by 1969, but by 1970, seventeen major regional systems spread across the country. Pendergrass, 48, 49; Rudolph and Ridley, 119.

[159] "Annual Project History, Region 2, Pacific Northwest-Pacific Southwest Intertie, Nevada-California," Vol I (1964), 11.

[160] "Annual Project History, Region 2, Pacific Northwest-Pacific Southwest Intertie, Nevada-California," Vol I (1964), 11-12; *Project Data*, 741, 743.

[161] "Annual Project History, Region 2, Pacific Northwest-Pacific Southwest Intertie, Nevada-California," Vol I (1964), 11-12; *Project Data*, 741, 743.

[162] Dominy, 63.

[163] United States Department of the Interior, Bureau of Reclamation, *Repayment of Reclamation Projects* (Washington, D.C.: Government Printing Office, 1972), 354.

[164] *Project Data*, 739-744.

[165] Emil V. Lindseth, "Researching the Intertie," *Reclamation Era* 51 (August 1965): 69.

[166] Lindseth, 69.

[167] IEEE defines high voltage as 72.5 kV to 242 kV (72,500 volts to 242,000 volts), and extra high voltage as between 242 kV and 800 kV (242,000 volts and 800,000 volts).

[168] Lindseth, 68-71.

[169] Newcomb Bennett, "A New Era of Power Transmission," *Reclamation Era* 51 (August 1965): 72-75.

[170] Thaddeus (Ted) Mermel, "D-C Developments in Other Countries," *Reclamation Era* 51 (August 1965): 86-89.

[171] Casazza, 28-33; White, 57, 59.

[172] East Central Area Reliability Coordination Agreement (ECAR); Electric Reliability Council of Texas (ERCOT); Florida Reliability Coordinating Council (FRCC); Mid-Atlantic Area Council (MAAC); Mid-America Interconnected Network, Inc. (MAIN); Mid-Continent Area Power Pool (MAPP); Northeast Power Coordinating Council (NPCC); Southeastern Electric Reliability Council (SERC); Southwest Power Pool (SPP); and Western Systems Coordinating Council (WSCC).

[173] Though Reclamation's Canyon Ferry Powerplant ties into the Mid-Continent Area Power Pool, the bureau does not participate in that reliability council.

[174] North American Electric Reliability Council, *NERC 1998 Annual Report*, 1, 3, 4, 24.

[175] Southwestern markets power in Arkansas, Missouri, Oklahoma, and Texas.

[176] *Federal Reclamation and Related Laws Annotated*, 801; *Annual Report of the Secretary of the Interior, 1944*, xii-xiii.

[177] Both Western and Southwestern market power in Kansas.

[178] Under the initial contracts, a small portion of power from the Pick-Sloan project is marketed in Minnesota and Iowa.

[179] Western does not market the energy generated at all of Reclamation's fifty-eight powerplants. Bonneville Power Administration markets energy generated at plants located in the Pacific Northwest.

[180] <http://www.wapa.gov/western/factsw.htm> accessed January 7, 2000.

[181] The Navajo Project is comprised of the Navajo Steam Generation Plant, a Southern Transmission System, and a Western (not DOE's) Transmission System. <http://www.wapa.gov/western/factsw.htm> accessed January 7, 2000.

"Glen Canyon Dam." Native American family overlooking Glen Canyon Dam. Colorado River Storage Project: Glen Canyon Unit, Arizona, Utah. Date: Unknown. Artist: Norman Rockwell.

CHAPTER NINE

Interaction between Reclamation and Native American tribes dates back to the early days of the organization. Reclamation conducted project investigations on reservations, often in conjunction with the Bureau of Indian Affairs,[182] and some of these projects were eventually constructed, either by Reclamation or one of the local Indian agencies. Generally, however, these interactions had to do with water supply as opposed to hydropower issues. In some cases, however, Reclamation did meet tribal energy needs by including hydropower production in its projects.

THE SACATON PROJECT

Reclamation included a transmission line to the Pima Indians' Reservation at Sacaton in the original 1909 plans for the Salt River Project transmission system in Arizona. Power delivery was part of a Federal agreement with the tribe to aid the pumping of well and irrigation water. In addition, Reclamation agreed to aid the Indian Irrigation Service of the Office of Indian Affairs in building a pumping plant on the Gila River Indian Reservation. The plant would then be controlled and operated by the tribe.[183]

The project became known as the Sacaton Project. Eventually, it expanded to include construction of a flood water canal and a line of wells and pumping plants connected by a ditch. The well and pumping plants were used to irrigate 10,000 acres on the north side of the reservation. In 1917, the Indian Irrigation Service revamped the project into the San Carlos Irrigation Project. At the same time, the Sacaton Indian School contracted with the Salt River Project to deliver non-irrigation power directly to the school.[184]

NAVAJO INDIAN IRRIGATION PROJECT

Another example of Reclamation constructing a project to benefit tribal lands was the Navajo Indian Irrigation Project (NIIP), a Colorado River Storage Project participating project. In 1962, Congress authorized a joint Reclamation and Bureau of Indian Affairs irrigation project to benefit the Navajo Reservation. Basically a network of irrigation facilities, the project used water from the Navajo Dam and Reservoir, the principal feature of the Navajo Unit of the CRSP. The project primarily served the residents of the Navajo Reservation in northeastern Arizona.

Initially, as part of the project, Reclamation planned to construct a small powerplant and switchyard at Navajo Dam to provide power for the pumping units required for irrigation. Initial site preparation began in the early 1970s. Concern for the proposed project's environmental impacts prompted the National Wildlife Federation and the New Mexico Wildlife Federation to file suit against the plant in 1976. A Federal judge ruled in 1977 that construction of the plant was not authorized by Congress. The ruling permanently halted

Aerial view of Navajo Dam and Reservoir. Colorado River Storage Project: Navajo Unit. Date: Unknown. Photographer: Unknown.

The Apache and her home. Salt River Project, Arizona. Date: Unknown. Photographer: Unknown.

Restored kiva or underground ceremonial chamber at Lost City (Pueblo Escondido), near St. Thomas, Nevada. Boulder Canyon Project, Arizona, Nevada. Date: May 13, 1934. Photographer: Ben D. Glaha.

Navajo women at ceremony. CRSP: Glen Canyon Unit, Arizona, Utah. Date: May 18, 1983. Photographer: Tom Fridmann.

construction on the powerplant.[185] The city of Farmington, New Mexico, later received a Federal Energy Regulatory Commission license and built a small powerplant at Navajo Dam. At the present time, electric power for NIIP is provided by CRSP.

NAVAJO GENERATING STATION

The Navajo Generating Station, located near rich coal deposits on the Navajo Reservation outside Page, Arizona, provided project power for Reclamation's Central Arizona Project and non-Federal purposes. The Navajo Tribe received energy from the project through transmission agreements. Additionally, the tribe received lease monies from the Peabody Coal Company, which mined coal on the reservation; this coal was then used to operate the generating station. The tribe negotiated "preferred employment" contracts with the Salt River Project, primary owner and operator of the plant. These contracts ensure jobs and job training for many Navajos.[186]

YELLOWTAIL DAM AND POWERPLANT

The highly ambitious Pick-Sloan Missouri Basin Program brought the Federal government and the local Native American tribes into conflict beginning with its authorization in 1944. Most of these battles occurred over land and water, not hydropower. Before construction of the Yellowtail Unit began in the 1960s, the Crow Tribe launched a court battle over right-of-way land for Yellowtail Dam and Reservoir. As part of their claims, the Crow Nation sought exclusive tribal use of the hydropower the project would generate. Congress repeatedly denied the tribe's request because it violated the clauses that gave public bodies, like municipalities or cooperatives, first preference to power.

In 1987, the Crow Tribe again requested exclusive use of hydroelectric power from the Yellowtail Unit, and Congress once again denied the request. In a compromise measure, Reclamation and the Crow agreed to enter into discussions to look at the feasibility of constructing a powerplant at the Yellowtail Afterbay Dam for tribal operation and use. The tribe felt the construction of a powerplant would help alleviate reservation unemployment and provide revenue. At the same time, a Reclamation study suggested that construction of a powerplant at the Yellowtail Afterbay Dam might help correct water quality concerns caused by the dam. Inability to gain political support for the project stalled and eventually halted these plans.[187]

An apprenticeship program, established at Yellowtail in the 1970s, proved relatively unsuccessful due to the remote location and the lack of local employees, causing a high employee turnover rate. Of the approximately 100 apprentices that completed the program between the

1970s and 1988, all but two transferred to other facilities. To address the turnover issue, Reclamation's Montana Area Office entered into an agreement in the late 1980s with the Crow Tribe to place tribal members into a tribe-sponsored apprenticeship program at Yellowtail. Only three of the eleven participants completed the tribal program due to hardships inherent to the new program. A second program, the Native American Apprenticeship Program, was subsequently established in which the participants were hired into Yellowtail's existing apprenticeship program. The overall Native American Apprenticeship Program proved highly successful in that five of the six Crow Tribal members that completed the programs were still employed at the facility years later. Largely due to its success in greatly reducing Yellowtail's employee turnover rate, the program ended in the mid-1990s.

Yellowtail Dam construction. Pick-Sloan Missouri Basin Program: Yellowtail Unit, Montana. Date: Unknown. Photographer: Unknown.

The Custer Substation near Custer, Montana, provided a vital link in the transmission of energy from the hydropower plant at Yellowtail Dam. Pick-Sloan Missouri Basin Program: Yellowtail Unit, Montana. Date: April 24, 1967. Photographer: Lyle C. Axthlem.

Bureau of Reclamation Commissioner Floyd Dominy and Mrs. Dominy upon arrival at the Hardin, Montana, airport for the dedication of Yellowtail Dam. Pick-Sloan Missouri Basin Program: Yellowtail Unit, Montana. Date: October 31, 1968. Photographer: Lyle C. Axthlem.

Yellowtail Dam. Pick-Sloan Missouri Basin Program: Yellowtail Unit, Montana. Date: July 12, 1967. Photographer: Lyle C. Axthlem.

General view looking along the axis of Yellowtail Dam from the right abutment. Pick-Soan Missouri Basin Program: Yellowtail Unit, Montana. Date: October 31, 1963. Photographer: T.R. Broderick.

GRAND COULEE DAM SETTLEMENT

In the early 1990s, the Confederated Tribes of the Colville Reservation filed suit against the United States[188] in the Court of Federal Claims. The tribes sought to recover damages under the "Fair and Honorable Dealings" clause of the Indian Claims Commission Act of 1946, alleging that the United States used land on the Colville Reservation to generate hydropower at Grand Coulee Dam and had promised to compensate the tribes for use of the land. The tribes had never received any remuneration.

After several years of negotiations, the tribes and the United States reached a settlement. The Bonneville Power Administration agreed to pay the Confederated Tribes fifty-three million dollars in a lump sum payment. Additional annual payments of a specified amount were to follow, beginning in fiscal year 1997. The agreement became binding after Congressional approval in late 1994. Although the Grand Coulee Dam Settlement only peripherally affected Reclamation, the issues addressed, specifically the unintended consequences of large-scale development of hydropower, remain important.

In 1989, Reclamation established an Office of Native American Affairs within the Commissioner's Office.[189] Under the direction of then Commissioner Dennis Underwood, the office grew to provide policy support for Reclamation, policy and technical support for Department of the Interior Indian water rights settlement efforts, and technical assistance when needed. The program continued to grow and take on new duties and responsibilities, including providing technical assistance to tribes evaluating their water and related resource needs and helping to develop practical solutions.[190]

Yakima at work in a hop vineyard, stripping the vines as she goes along. Yakima Project, Washington. Date: September 1907. Photographer: Walter J. Lubken.

Footnotes for Chapter 9: Reclamation's Power Program and Native Americans

[182] Reclamation projects constructed for tribal use included, among others, the Blackfeet Indian Irrigation Project, Fort Peck Indian Project, San Carlos Project, the Flathead Project, and Fort Hall Indian Project.

[183] F.H. Newell, *Third Annual Report of Reclamation Service 1903-4* (Washington, D.C.: Government Printing Office, 1905), 62, 152, 156.

[184] Pfaff, *Theodore Roosevelt Dam, Power Plant*, 14, 16; Arthur Powell Davis, *Annual Report of the Reclamation Service, Department of the Interior* (Washington, D.C.: Government Printing Office, 1917), 25; Christine Pfaff, *San Carlos Irrigation Project* (National Park Service, Historic American Engineering Record, No. AZ-50), 13-15.

[185] Leah S. Glaser, *The Navajo Indian Irrigation Project* (Denver: Bureau of Reclamation History Program), 1998; Toni Rae Linenberger, *The Navajo Unit, Colorado River Storage Project* (Denver: Bureau of Reclamation History Program), 1998.

[186] Arizona Public Service, a private utility, negotiated a power agreement where the Navajos would receive a significant electrical supply from the Four Corners Power Plant. Jo Heyer, "Progress through Power: Electrical Training for the Navajo Tribe" (Santa Fe: Reynolds Electrical & Engineering Co., 1962), 51; "Power Facilities to Become Reality," *Navajo Times* (December 1960), 1; Linenberger, *The Navajo Unit*, 32.

[187] Project History, "Annual Project History, Yellowtail Unit, Montana, Wyoming" (1963), 27-31; Lawson, 183-184; *Project Data*, 1013-1016.

[188] The lawsuit specifically cited the Bonneville Power Administration and the Department of Energy. Reclamation was not part of either the lawsuit or the negotiated settlement, except as the producers of energy marketed by BPA.

[189] Previously, Reclamation dealt with Native American issues in the context of water rights claims.

[190] <http://www.usbr.gov/main/programs/native-am.html> accessed August 26, 1998.

The setting sun over Lake Lowell. Boise Project: Arrowrock Division, Idaho. Date: December 3, 1975. Photographer: Glade Walker.

CHAPTER TEN

Harnessing rivers inevitably affects the environment. However, most people did not begin to fully understand many of the issues involved until after World War II. During the 1960s and 1970s, the environmental movement expressed the public's growing concerns through traditional routes, such as newspapers, books, and articles, and through not-so-traditional routes, such as popular music, film, and television. Increased public consciousness and support for environmental causes manifested itself in political action. This political action precipitated additional research studies of dams, irrigation, hydroelectricity, and the environment. All this public and political attention gradually prompted increased environmental consciousness on the part of government water resource managers.

CHANGING PERCEPTIONS ABOUT THE ENVIRONMENT

Dating back to the early decades of the twentieth century, electricity stimulated tremendous growth and opportunity across the West.[191] Proponents viewed the reclamation process as natural and desirable for the West and technology as beneficial to livability and development. In his book about the Columbia River, entitled *The Organic Machine*, Historian Richard White observed that, "In the 1930s, proponents of dams were more likely to see dams as mimicking nature rather than conquering it."[192] Conservation issues were an early concern for Reclamation, crucial to passage of the Reclamation Act in 1902 and the subsequent development of hydroelectric power. Even so, early conservationists, such as Theodore Roosevelt and Gifford Pinchot, focused on maximizing the use of natural resources for the maximum number of people (the greatest good for the greatest number of people) while avoiding waste.

At the same time, others expressed concern over the effect of water storage and hydroelectric development on the environment. Between 1908 and 1912, the public supported the conservation of Lake Tahoe's natural lake level and scenic beauty, when Reclamation proposed to build a dam to increase storage capacity and to sometimes lower the existing lake level to benefit the Newlands Project. Subsequently, proposals for Reclamation projects raised consciousness about the impacts of major dams on various resources. During President Truman's Administration, several preservation-minded groups accused Reclamation of ignoring the aesthetic impact of its projects. The preservationists' concerns with aesthetic impacts ultimately evolved into more specific issues like water quality, air quality, and sustainable ecosystems.[193]

Negative publicity occurred for Reclamation when its projects intruded upon publicly enjoyed and recognized areas like national parks. For years, Coloradans and their Congressional representatives pushed Reclamation to build the Colorado-Big Thompson Project, which would require construction on the fringe of and under

William A. Simmonds, gate tender at Lake Tahoe Dam. Newlands Project, Nevada, California. Date: Unknown. Photographer: Unknown.

Workmen in the process of placing steel supports in the gate chamber area of the Alva B. Adams Tunnel. Colorado-Big Thompson Project, Colorado. Date: Unknown. Photographer: Unknown.

Rocky Mountain National Park. The project was ultimately built because the park's enabling legislation specifically authorized a water development project infringing on its boundaries. In the 1950s, the controversy over proposed construction of Echo Park Dam in Dinosaur National Monument in Colorado heightened public awareness of issues surrounding construction of a dam in a National Park Service managed area. Ultimately, public opinion forced cancellation of plans for Echo Park Dam and resulted in construction of the alternative, Glen Canyon Dam.

In the 1960s, Reclamation engineers suggested construction of Marble Canyon and Bridge Canyon, both run-of-the-river hydroelectric dams just outside the boundaries of Grand Canyon National Park. Secretary of the Interior Stewart Udall canceled these proposed hydropower projects due to public outcry and pressure from environmental groups and Native American tribes. Surprisingly, opponents of these two dams sided with nuclear power proponents against the hydropower projects. The environmental issues related to nuclear powerplants had not yet become embedded in America's consciousness.

The change in the way the majority of Americans viewed the exploitation of natural resources in the American West resulted in part from improved communication technologies during the 1950s and 1960s. News came not just from newsreels, radio, and newspapers, but from television, a medium that included actual, immediate images. Advances in transportation meant that the average American could travel to "the West" on airliners or in powerful new automobiles on improved highways. Once there, visitors came to a more sophisticated understanding about the issues in the West and began to consider the region a part of the American identity. Thus, large new groups of Americans gained both a feeling of ownership and responsibility toward public lands and public works, many of which lie in the West. American support for regulation of exploitation of historic and natural resources grew as heightened awareness of the effects of technology on these resources rose.[194]

As public awareness of technology and support for management of natural and historic resources increased, so did the search for ways to address growing public concerns. Not surprisingly, initial attentions turned toward technology to solve existing problems. While technology often caused the initial problem, it also promised the solution.

Glen Canyon Dam. Colorado River Storage Project: Glen Canyon Unit, Arizona, Utah. Date: May 17, 1974. Photographer: Unknown.

ENVIRONMENTAL OPPOSITION TO RECLAMATION PROJECTS

In the late 1940s and early 1950s, nuclear energy was one of the technological competitors to traditional hydroelectric powerplants and extractive coal-, oil-, and natural gas-fired steam generating plants. Nuclear advocates argued that nuclear powerplants generated more kilowatts per plant than traditional plants, thus making them more economical. In addition, the atom reduced dependence on rapidly depleting natural resources. At the same time, the nuclear industry and its supporters played to Americans' faith in technology and general fascination with all things new by touting developments in atomic energy. As environmental awareness continued to develop among the general populace, the nuclear industry found another unexpected avenue of support. In widespread publicity, nuclear proponents within and outside the industry claimed that nuclear power could replace and even expand existing hydro and steam powerplant capacity without any of the environmental repercussions.

The lure of an easy solution to many existing environmental problems attracted many, who campaigned in favor of building nuclear powerplants instead of new hydropower facilities, which would require new dams and reservoirs. Media support, pointing out the long-lasting environmental and aesthetic repercussions of river manipulation, convinced a wider audience of the validity of the argument. The public's belief that nuclear power generation provided a viable option to meet growing electric power needs in the West fueled opposition to the construction of various Reclamation hydroelectric projects.

One of the most notable examples of the conservation movement siding with nuclear advocates to oppose selected construction remains the Sierra Club's objection to Reclamation's proposed dam at Echo Park in Dinosaur National Monument.[195] The Sierra Club worked in conjunction with other wilderness and conservation organizations to contest the project. The fight to save Echo Park firmly established the Sierra Club as a national environmental organization, as well as a force to be reckoned with in legislative affairs.

Reclamation proposed Echo Park Dam as part of the Colorado River Storage Project plan to develop the Upper Colorado River Basin. According to Reclamation engineers and planners, Echo Park was a key feature of the Central Utah Project[196] portion of CRSP. Led by David Brower, Executive Director of the Sierra Club, conservationists opposed a dam in the national monument because it would infringe upon the scenic beauty and violate the sanctity of the monument. Instead, Brower and the Sierra Club supported and advocated for a high dam at Glen Canyon, arguing the increased storage in Glen Canyon Reservoir (Lake Powell) provided as much storage as a dam in Echo Park would, with comparable evaporation.

Indian petroglyphs at Atlatl Rock in the Valley of Fire near St. Thomas, Nevada, at about the time the area became a Nevada State Park. Date: April 20, 1935. Photographer: Ben D. Glaha.

The "atoms-not-dams" argument made by Brower and Club members, in conjunction with an attack on Reclamation's financial and evaporation figures for the project, served to sway Secretarial and Congressional support away from construction of the Dinosaur portions of the project. The deletion of the dam at Echo Park eventually resulted in construction of the proposed dam at Glen Canyon, a compromise the Sierra Club supported.[197] Furthermore, this fight to save Echo Park occurred in the 1950s, even before the so-called, "environmental revolution" of the 1960s.

Commissioner Ellis Armstrong (at right) talks with H.R. Gnau, Chief, Glen Canyon Field Division. Colorado River Storage Project: Glen Canyon Unit, Arizona, Utah.
Date: May 1972.
Photographer: Stan Rasmussen.

At the time, a high dam at Glen Canyon was thought to be a better alternative to the dam at Echo Park, largely because Glen Canyon did not directly impinge upon a National Park Service managed resource. Reclamation assured conservationists that they would construct a dike to prevent flooding of Rainbow Bridge National Monument by the reservoir; this proposal was not well received by all of Reclamation's leadership. In an oral history interview, Floyd Dominy said of plans for the dike:

> I was flabbergasted. Here is a remote canyon country with no access whatever, and you'd have to put a barrier dam below the monument. You'd have to go above the monument, put in another barrier dam. [To prevent runoff from pooling behind the barrier dam designed to keep Lake Powell from backing up into the monument.] You'd have to drill a tunnel from one canyon over to another, about two miles. And to make the estimate of what kind of volume of water you're talking about, what comes off of Navajo Mountain in a thunderstorm, what comes off when the snow melts, why, the violence to that scenic grandeur to perform this kind of construction would be so goddamn much more horrendous than to let a little water back in under that bridge.
>
> You're not going to get up on the arch of the bridge. There's an arroyo 85 feet deep from the bridge abutments, and we're only going to put fifty-four feet of water in that arroyo. And there's a spring coming out of one abutment, so you know that water saturation isn't going to hurt anything. Anybody could tell that. You didn't have to have geologists. I got geologists to confirm that opinion later.[198]

Reclamation later scrapped plans for the dike, claiming that added construction would scar the landscape and detract from the scenery around the bridge.

As Congress prepared to pass the CRSP Act, Brower discovered the scenic beauty of Glen Canyon and attempted to persuade Sierra Club members to oppose construction of a dam. The board declined Brower's request because they felt that any further protests against CRSP weakened their case against Echo Park and was outside the scope of current Sierra Club goals. In addition, taking a stand against Glen Canyon left the club open to censure for attempting to stop development in the Upper Colorado River Basin and accusations of lobbying for southern California.[199]

The environmental-nuclear alliance continued into the 1960s, when Reclamation proposed a set of dams at the Grand Canyon (one at the northern edge and one at the southern) just outside the boundaries of Grand Canyon National Park. Reclamation designed Bridge and Marble Canyon Dams to provide hydroelectric power to pump water on the Central Arizona Project. These dams were originally proposed in the late 1940s, but when the idea for the dams resurfaced in the late 1950s and became part of the Pacific Southwest Water Plan in 1964, conservationists felt the Bridge and Marble Canyon Dams would intrude upon the national park. Despite Reclamation assurances that the dams and most of the reservoirs would remain outside the park and not spoil the view of the canyon conservationists revolted. The battle for Bridge and Marble Canyon Dams far surpassed that of Echo Park Dam. Environmental groups, nuclear power proponents, and Native American tribes all joined forces to fight construction of the Grand Canyon dams.

When Reclamation and Floyd Dominy proposed the Grand Canyon dams, they did not realize the impact the proposal would have on the American public and the world. To many people, the Grand Canyon was considered a sacred site; Reclamation was not proposing to dam only a geographic site, it was proposing to dam a mythic one as well. Led by David Brower and the Sierra Club, those opposing the dams capitalized on national affinity for America's own "natural wonder of the world." Using both traditional – newspapers, books, and articles – and not-so-traditional media, as well as film, popular music, and television, Brower waged an all-out war against Reclamation and the dams. Not to be outdone, Dominy retaliated in much the same fashion. Both sides published numerous articles, pamphlets, and books; gave interviews; and generally tried to sway public opinion.

The Sierra Club's efforts culminated with a full-page ad on the inside front pages of the *New York Times*, the *Washington Post*, and the *Los Angeles Times*, saying, "*Now only you can save the Grand Canyon from being flooded . . . for profit.*" This ad received considerable attention, most notably from the Internal Revenue Service (IRS), "which revoked the Sierra Club's tax-exempt status for engaging in 'substantial' efforts to

Looking upstream at Bridge Canyon damsite on the Colorado River. Date: Unknown. Photographer: Unknown.

Mile 39.5 in Marble Canyon on the Colorado River in northern Arizona was one of the most inaccessible damsites ever explored by Reclamation engineers. Marble Canyon Dam, if built, would have spanned the river between the sheer inner gorge walls in the foreground. Date: August 1951. Photographer: John Santa.

influence legislation."[200] In a single move, the IRS moved the issue of damming the Grand Canyon from the inside cover to the front page. Public response was immediate and considerable, almost all of it in favor of the Sierra Club's position. Riding the wave of public support, and in response to Dominy's argument that the dams would increase tourism to the area, Brower developed another full-page ad. This one asked, "*Should we also flood the Sistine Chapel so tourists can get nearer the ceiling?*" Again, the response came back as a resounding "NO." The public flooded Congress with hundreds of thousands of letters, emphasizing this point.[201]

In 1967, Secretary of the Interior Stewart Udall finally canceled the projects due to these outside pressures. In a compromise, in which neither the environmentalists nor the nuclear industry prevailed, Reclamation received a 24.3 percent entitlement to power that was generated at the coal-fired, non-Federal Navajo Steam Generating Station on the Navajo Reservation near Page, Arizona.[202] Ironically, coal-fired power generation had its own environmental effects, most notably on the air quality in the immediate area.[203]

Environmental Impacts of Hydropower Plants

As the environmental movement gained strength and the public became more aware of widespread environmental and reliability issues, Reclamation began looking at the various modifications, both positive and negative, resulting from its projects. Independently, hydroelectric powerplants possessed many environmentally friendly characteristics. They produced virtually no pollution, consumed no non-renewable natural resources, and required smaller amounts of labor. Hydroelectricity could be generated twenty-four hours a day, and some sites had tremendous capacity for expansion. In 1990, the Federal Energy Regulatory Commission estimated the undeveloped hydropower potential in the United States was double that of current capacity.[204]

Those who were environmentally aware quickly pointed out that hydropower plants did not act independently, however. Outside of a few run-of-the-river facilities, hydropower plants required storage dams that dramatically altered the natural environment. The side benefits of hydropower production and storage dams – increased agriculture and urban growth – alter the environment as well. When settling in new places, humans altered the environment for agriculture, irrigation, or settlement. In the name of progress and development, humans cut down trees, diverted water, realigned natural growth patterns, and built on flood plains and in other undeveloped areas. Neither man nor nature exists in a void. An event in one place had the power to affect an entire system, and eventually effects could be and often were widespread. For example, the collapse of the salmon fisheries in the Pacific Northwest, and the resulting cultural and economic implications, has been attributed to the ecological changes wrought by dam construction and development of the Columbia River System.

Any method of generating power caused unique environmental impacts. Coal and natural gas plants polluted the air. Strip mining for coal destroyed the landscape, oftentimes far from the actual plant. Nuclear powerplants potentially could emit harmful radiation into the air and water and onto the land. Nuclear waste caused disposal problems because it remained dangerous for centuries. Hydropower plants modified the flow of rivers, changing the annual hydrograph and water quality. Rivers were turned to lakes, the ecology changed, populations of aquatic organisims changed, and fish movement up and down river was restricted.

Water discharged through powerplants created turbidity in previously still waters. Turbulent waters, in turn, changed the concentration of various gases, including nitrogen and oxygen, in the water. These gases caused various maladies in fish and encouraged growth of different organisms and plants in the water. In the case of pump storage (also known as pump-generating) hydropower plants,[205] the regulated releases caused even more drastic effects. The daily rise and fall of reservoir water levels further affected the wetland and riparian environment along the shoreline.

Even without the supplemental benefits of hydropower production, the practice of storing and releasing irrigation water changed natural river flows. Dams blocked traditional spawning grounds for fish and other native species. The resulting reservoirs flooded additional breeding grounds and river habitat. At the same time, however, new reservoirs offered new recreational opportunities to local communities.

Releases from reservoirs through dams also changed the ecology of the river. Water releases from the bottom of reservoirs were often colder than the previously free-flowing river. Some fish thrived in the colder river below dams, while others, traditionally the native species, did not. These introduced fish and the colder temperatures drove out the native fish populations, altering the natural ecosystem of the river. At the same time, the thriving non-native fish populations often provided sport fisheries of the highest caliber, drawing anglers from around the world.

Native Americans fishing at Celilo Falls, Columbia River, near The Dalles, Oregon. Date: 1944. Photographer: Ben D. Glaha.

Later, Reclamation attempted to control these drastic temperature variations by installing temperature

Aerial view of a large part of Hanford's 200 Area, including the liquid effluent treatment facility, effluent treatment facility, Plutonium-Uranium Extraction Plant, and a powerplant. Date: Unknown. Photograph courtesy of the Department of Energy.

control devices on several of its facilities. One such device was installed on Shasta Dam in the 1990s. The primary purpose of the temperature control device was to manage cold water resources by allowing for selective withdrawal of water from four different reservoir elevations. Conservation of the cold water pool was achieved by forcing water withdrawal from the highest elevation possible, while meeting downstream temperature targets. In 2002, the temperature control device was selected as a recipient of the National Hydropower Association's award in the technical solutions category.

Anglers were affected by variations in river flow based on irrigation needs, as was hydropower production. On Reclamation projects, irrigation affected the generation of power. Initially, water only ran through generators during the late summer irrigation season, when Reclamation made water deliveries to water users or during high spring runoffs, when additional water storage was unnecessary. With the construction of Hoover Dam, that pattern changed slightly, but power production still remained dependent upon the regulated releases for a variety of project purposes from the storage dams on a river.

ENVIRONMENTAL LEGISLATION

People began to notice this changed environment as they learned more about the natural world. The activists among them elevated the issues to a national level, making them a part of the public consciousness. At the same time that environmental issues were becoming part of the national consciousness, they also naturally entered political dialogue. At the prompting of Senator Gaylord Nelson, in the fall of 1963, President Kennedy launched a tour to give national visibility to environmental issues. Unfortunately, the President's tour yielded few results and did not make the environment a national political issue. On a conservation speaking tour in the West six years later, in July of 1969, Senator Nelson observed protests against the Vietnam War on college campuses. After contacting governors, mayors, newspapers, and magazines across the country, he initiated Earth Day to mimic such grassroots activism. The one-day national event in the spring of 1970 brought the

Aerial view of temperature control device at Shasta Dam. Central Valley Project: Shasta/Trinity Division, California. Date: Unknown. Photographer: Unknown.

environment into the public dialogue with dramatic success on both local and national levels.[206]

The public emphasis on the environment brought widespread awareness of the impact of man on his natural surroundings. In turn, people began to realize land reclamation, water project development, and dam construction caused ecological changes that altered plant, fish, animal, and bird populations through displacement or destruction. In addition, projects often destroyed historic or archeological resources. Simply tearing down the dams would likely not remedy the damage. Instead, to address the ecological impacts of human activities on the environment, Congress passed a flurry of environmental laws between 1964 and 1978.[207] These included:

- The Wilderness Act (1964)
- The Anadromous Fish Conservation Act (1965)
- The National Historic Preservation Act (1966)
- The National Environmental Policy Act (NEPA) of 1969
- The Environmental Quality Improvement Act of 1970
- The National Inspection Program (1972)
- The Clean Water Act (1972)
- The Endangered Species Act of 1973
- The Federal Land Policy Management Act of 1976
- Reclamation Safety of Dams Act of 1978

Presidential Executive Orders, Federal regulations, and state and local laws accompanied these laws. Not surprisingly, these new laws and regulations affected both public and private hydropower programs.

In response to environmental impact study requirements, new legislation, and public protests, Reclamation began changing the way it did business, as did many other public and private utilities. Beginning in the early 1970s,

Fishing below Jackson Lake Dam. Minidoka Project, Idaho. Date: 1918. Photographer: J.C.C.

The public campground at Barretts Diversion Dam. Pick-Sloan Missouri Basin Program: East Bench Unit, Montana. Date: July 20, 1966. Photographer: Lyle C. Axthlem.

Easton Dam and fishway. Yakima Project, Washington. Date: May 24, 1932. Photographer: George A. Beyer.

Steelhead clear ladders during one of the largest spawning runs on record in Oregon. Date: June 27, 1997. Photographer: David Walsh.

Reclamation built centralized staffs and programs to deal directly with growing environmental and cultural resource issues. Previously, Reclamation dealt with environmental and cultural resource issues on each individual project at the project office level.[208]

ENDANGERED SPECIES

Dams have affected fish populations throughout the West. Through years of operating history, the public and Reclamation have learned more about the extent to which dams can alter the environment and affect fish and wildlife species. Releases from reservoirs can change the water temperature and the level of dissolved gases such as oxygen and nitrogen. Reservoirs also trap sediments which may contain and concentrate potentially toxic materials from upstream sources. Power operations can cause water level fluctuations that affect downstream riparian areas.

Dams have blocked fish migration and eliminated access to spawning and rearing habitat. The reservoirs slow the juvenile fish migrations to the ocean, which increases their exposure to disease and predation. Juvenile fish (smolts) also suffer mortality when migrating downstream through dams via the turbines. The most publicized instance of these environmental problems associated with dams remains the salmon[209] in the Northwest due to widespread construction activities throughout the river basin by Reclamation and the Corps, and several public and private utilities.[210]

Grand Coulee Dam blocked 1,100 miles of river when built. Similar losses occurred during construction of private developments in the 1960s. In 1939, Reclamation acknowledged that fish could not successfully overcome the height of Grand Coulee Dam and, with the guidance of fishery management organizations, began a program to capture migrating fish below Grand Coulee and transport them to spawning streams above the dam for release. By the end of the six-year program, Reclamation "anticipated that the migratory fish will be reestablished, probably with the result that runs will be increased."[211] Unfortunately, the experiment was not successful, and project impacts were mitigated with a hatchery program. Since 1939, five public utility and four Corps dams were built on the Columbia. Though these dams all included fish ladders to aid fish passage upstream, their construction seriously compounded the problem.[212]

The series of dams up the Columbia and lower Snake Rivers, constructed by the Corps, allowed barge navigation as far inland as Lewiston, Idaho, and provided hydropower benefits. While a boon to the Northwest economy, the dams made salmon migration and spawning difficult. In the mid-1990s, various environmental groups proposed breaching or dismantling a series of four dams on the Snake River, the largest tributary of the Columbia. The measure received considerable attention, both publicly and politically, and was the subject of a study by the Corps. In December of 2000, the

National Marine Fisheries Service issued a biological opinion on the effect of the hydropower system on twelve threatened and endangered salmon and steelhead stocks. The opinion listed a host of actions to take to recover the fish. Dam removal was considered but not included.

Salmon issues were not limited to the Columbia/Snake River system. The San Joaquin/Sacramento River system in northern California was similarly impacted. As with the salmon recovery efforts in the Pacific Northwest, Reclamation, in cooperation with other private, state, and Federal entities, undertook mitigation efforts to help sustain the species. The long-term success of this effort remains to be seen.

Fishing on Strawberry Reservoir. Strawberry Valley Project, Utah. Date: Unknown. Photographer: Unknown.

Footnotes for Chapter 10: Reclamation and the Environment

[191] Kathka, 28-29.

[192] White, 71-73; Casazza, 27.

[193] Kathka, 138-140.

[194] Brit Storey, "Brief History of the Bureau of Reclamation."

[195] Early conservationists also believed that natural resources should be protected for public enjoyment. To these ends, lawmakers established the National Forest Service (1891), the National Park Service (1916), the Reclamation Service (1902), and various national parks and monuments. Lawmakers and a large portion of the public maintained that by preserving representative parcels of land, the remainder could be utilized without significant loss to the public at large.

By the 1950s, the definition of conservation had changed to emphasize the preservation portion of the philosophy. These new conservationists saw the purpose of the national park system as the protection of natural resources. According to these activists, national parks did not exist merely for public enjoyment, they protected our natural resources from development. The original concept of managed use gave way to preservation of wilderness and finally environmentalism. The 1950s and 1960s reflect the gradual shift from early Progressive definitions of conservation, efficient management for the good of the many, to the more modern environmentalism and looking at man's effect on the natural world.

See also Mark W. T. Harvey, *A Symbol of Wilderness: Echo Park and the American Conservation Movement* (Albuquerque: University of New Mexico Press, 1994); Jared Farmer, *Glen Canyon Dammed: Inventing Lake Powell and the Canyon Country* (Tucson: University of Arizona Press, 1999), 137; Samuel P. Hays, *Conservation and the Gospel of Efficiency: The Progressive Conservation Movement, 1890-1920* (Cambridge: Harvard University Press, 1959).

[196] The Central Utah Project is one of the participating projects in the Colorado River Storage Project.

[197] Thomas Raymond Wellock, *Critical Masses: Opposition to Nuclear Power in California, 1958-1978* (Madison: University of Wisconsin Press, 1998), 41; for more information on the fight to save Echo Park, see Harvey.

[198] Oral history interview, Brit Allan Storey of Floyd E. Dominy, April 8, 1996, Boyce, Virginia.

[199] See Harvey; Wellock.

[200] Farmer, 152.

[201] Farmer, 152-153; Philip L. Fradkin, *A River No More: The Colorado River and the West* (New York: Alfred A. Knopf, 1981), 228-234.

[202] The other owners include the Salt River Project in Arizona, the Los Angeles Department of Water and Power, the Arizona Public Service Company, Nevada Power Company, and Tucson Electric Power. <www.glencanyon.net/navajo.htm> accessed September 1, 1998; Brit Storey, "Brief History of the Bureau of Reclamation."

[203] Brit Storey, "Brief History of the Bureau of Reclamation"; Harvey, 62.

[204] United States Department of the Interior, Bureau of Reclamation, *Reclamation's Energy Initiatives: Taking a Leadership Role in Supporting the National Energy Strategy* Draft (October 1990), 7.

[205] Pump storage plants are designed to take advantage of peaking power needs. A pump storage plant uses water from a reservoir upstream to generate power when demand is high, usually during the daytime hours. After water passes through the generators, it is captured in a downstream reservoir. At night, when the demand for power is less, water is pumped from the lower reservoir back into the upper reservoir. By pumping water when demand is low and generating power when demand is high, the income from the sale of power is greater than the costs to run the pumps, making the entire operation economically viable.

[206] Gaylord Nelson (Senator), "History of Earth Day," <www.earthday.org> accessed August 20, 1998.

[207] Klotz, 60.

[208] For cultural resources, generally when a project required mitigation work, Reclamation contracted with the National Park Service to perform any necessary investigations. Reclamation's actual involvement in cultural resources dates to at least the 1940s, when they helped fund the River Basin Surveys conducted by the Smithsonian Institution.

[209] Twelve species of salmon and steelhead – mostly found in Oregon, Idaho, and Washington – are listed as threatened or endangered in the Columbia River Basin. Several others are listed in California.

[210] Robert Goodland, "Environmental Sustainability in the Hydro Industry: Disaggregating the Debate," in Tony Dorcey, *Large Dams: Learning From the Past Looking at the Future* (Gland, Switzerland: The World Conservation Union, April 1997), 90-96; Lisa Mighetto, "Salmon, Science, and Politics: Writing History for the U.S. Army Corps of Engineers," *The Public Historian* 17 (Fall 1995): 22-24. See also <www.salmonrecovery.gov>.

[211] United States Department of the Interior, *Annual Report of the Secretary of the Interior* (Washington, D.C.: Government Printing Office, 1939), 196.

[212] White, 89, 95; Pitzer, Paper, Western History Association, 9.

At sunrise on March 26, 1996, Secretary Bruce Babbitt turned a small wheel, and as millions of people watched on national TV, a rushing plume of water shot forth, spanning nearly a quarter-mile of the canyon below Glen Canyon Dam. Close-up view of hollow jets. Colorado River Storage Project: Glen Canyon Unit. Date: March 1996. Photographer: Tom Smart. Photograph courtesy of *Deseret News.*

For seven days in March of 1996, the outlet works at Glen Canyon were opened to simulate a natural flood to rebuild beach and habitat below the dam. Photo of the dye used in the flooding. Colorado River Storage Project: Glen Canyon Unit. Date: March 1996. Photographer: Tom Smart. Photograph courtesy of *Deseret News.*

View of the powerplant at American Falls Dam. Date: May 23, 1984. Photographer: Glade Walker.

Fishing at Pueblo Dam. Fryingpan-Arkansas Project, Colorado. Date: Unknown. Photographer: Unknown.

Lower Molina Powerplant. Collbran Project, Colorado. Date: June 10, 1998. Photographer: Bradford P. Osburn.

Due to higher than normal snowpack, excess water was released over the spillways at Hoover Dam. Arizona spillway shownBoulder Canyon Project, Arizona, Nevada. Date: 1983. Photographer: Gene Hertzog.

Aerial view of Twin Lakes Reservoir showing the Mount Elbert Pumped-storage Powerplant in the center of the photo on the far shore. The Mount Elbert forebay and dam are above the powerplant in the top center of the photo. Fryingpan-Arkansas Project, Colorado. Date: August 15, 1986. Photographer: Donald Westphal.

CHAPTER ELEVEN

The post-World War II boom in energy consumption came to an abrupt end in the early 1970s, when the Organization of Petroleum Exporting Countries (OPEC) raised the price of oil to the United States.[213] For a long time, the benefits of fossil fuels were considered superior to renewable sources, and fossil fuels were heavily used to produce energy. However, during the Arab oil embargos from 1973 to 1974, and in the following years, the oil crisis shifted American opinion in support of renewable energy.[214] The shortage highlighted the benefits of hydropower, which does not consume water, but actually uses falling water to create power. In this way, it is considered a clean and renewable energy source. Other examples of renewable energy sources are wave power, tidal power, wind power, and solar power.[215]

With President Jimmy Carter's endorsement, Congress passed the Public Utility Regulatory Policy Act (PURPA) in 1978 to help alleviate the energy crisis and ensure protection in times of gas shortages. Its passage encouraged the use of alternative energy sources and promoted conservation.

In 1979, Secretary of the Interior Cecil Andrus changed Reclamation's name from the Bureau of Reclamation to the Water and Power Resources Service. The new name attempted to reflect the bureau's new mandate as a water management bureau instead of a dam-building organization. It also highlighted the importance of the power program within the bureau and promoted the renewable energy sources utilized by Reclamation. The new name was not well received, however, and in 1981, Secretary James Watt restored the name "Bureau of Reclamation." Still, the internal and external emphasis on alternative and renewable energy sources continued.

As scientists tested and developed new renewable energy sources, people began to recognize the benefits. Recent developments in and better building materials for machines that harness renewable energy sources led to higher outputs of energy and helped bring down the price of renewable electricity sources. As the price decreased, renewable energy sources became more economical and, subsequently, a more popular source of energy. In the late 1970s, Reclamation constructed an experimental wind farm near Medicine Bow, Wyoming. Designed to capitalize on the high winds in the area, electricity generated at the wind farm was tied into the CRSP transmission system. Over time, Reclamation discontinued its involvement in wind technology.

UPRATING

In addition to taking advantage of renewable energy sources like wind, Reclamation investigated specific technology called "uprating" and "rewinding" to improve the efficiency of its hydropower facilities. Faced with a dwindling supply of natural resources, political and public antipathy to new construction projects,

Giant wind turbines have been installed in southwestern Washington state. The Bonneville Power Administration and Washington Winds Inc. are developing a 150 mW wind farm that will generate enough electricity on average to serve more than 36,000 homes. The Maiden Wind Farm would be located about fifteen miles north of Prosser, Washington, in Benton and Yakima Counties. Date: June 15, 2001. Photographer: David Walsh.

Working on Unit A-9 at Hoover Dam. Boulder Canyon Project, Arizona, Nevada. Date: November 5, 1990. Photographer: Carl Trulsen.

Generator thrust bearing and upper guide bearing bracket lift on Unit No. 4 at the start of uprating activities at Shasta Powerplant. Central Valley Project: Shasta/Trinity River Division. Date: 1997. Photographer: Unknown.

and continued increasing demand, the electric industry began to recognize uprating as one of the better long-range solutions to help alleviate the energy crisis. Uprating allowed existing hydropower plants to capitalize on available water resources and bring additional capacity into the system. As a result, Reclamation began a program to uprate many of its older generators.

Issued in 1977, the Western Energy Expansion Study identified several proposals for uprating existing Reclamation hydroelectric units. The study concluded that uprating existing units would be the most immediate, cost-effective, and acceptable means of contributing to the Nation's generating capacity. In addition, a Government Accounting Office audit and study[216] recommended that Reclamation evaluate opportunities to increase production and subsequently act on the economically justified opportunities. Following these recommendations, Reclamation and Interior embarked on a program in 1978 to investigate and implement all viable options for improving existing plants by modernizing and uprating the generator equipment.

Beginning with two units located at Shasta, California, Reclamation launched an uprating program. To increase the energy capacity at its facilities, Reclamation began uprating as funding and unit availability allowed. Reclamation rewound[217] the generators, updated or replaced associated equipment, and occasionally replaced turbines. Newer turbines have a two to four percent higher efficiency. Replacing the older turbines increased the efficiency of the plant, as well as increased the capacity of the unit. At the same time, Reclamation rewound a number of generators that had windings in poor condition, but where no appreciable uprate potential existed.

Reclamation entered into an uprating program where it studied the capability and limits of all power equipment from the penstock through the turbine, generator, bus, switchgear, transformer, and transmission system. It uprated units where power capacity increases greater than fifteen percent could be obtained. Uprating was possible because many of Reclamation's units had bigger turbines than necessary so they could operate efficiently at lower reservoir levels. It turned out that Reclamation reservoirs tended to operate at higher levels than originally planned for in the generator design. Because the generators were designed for lower reservoir levels, they were unable to generate to the full potential of the turbines when reservoirs were operated at higher levels. By uprating, generators could operate to the full potential of the turbines when reservoirs were operated at higher levels. Advances in insulation technology allowed Reclamation to rewind the generator units and obtain a higher power capacity rating.

The most significant factor in determining if a generator should be uprated was economics. A demand must have existed for the additional capacity provided by an uprate before Reclamation considered the project economically feasible. Environmental concerns also played a role in determining project viability. If the environmental impacts of an uprate outweighed its necessity, the project was taken out of consideration.

By 1999, Reclamation completed uprating and rewinding units at thirty-three of its fifty-eight powerplants, or roughly three-quarters of the units in operation. Over the course of the uprating program, Reclamation's generating capacity increased by 1,632,823 kilowatts. The capacity of the uprated units increased an average of fifty percent, representing more than eleven percent of Reclamation's total power generating capacity.[218]

RE-OPERATION OF GLEN CANYON DAM

In the late 1980s and early 1990s, environmental concerns about the effect of dams on ecosystems reached an all-time high. This greater environmental awareness brought with it a desire for improvement.

In July of 1989, the Secretary of the Interior directed Reclamation to prepare an environmental impact statement (EIS) evaluating the effects of Glen Canyon Dam operations on downstream resources, specifically the impact on the ecological and cultural resources of Glen Canyon National Recreation Area and Grand Canyon National Park. With the cooperation of several other entities,[219] Reclamation filed the final EIS on March 21, 1995. The EIS attempted to determine how the change in the historic river flow – from the wild, sediment-filled, flood-prone Colorado River of old to the dammed and managed river of the late twentieth century – affected the downstream environment. Construction of the dam and the formation of Lake Powell in the late 1960s altered the river environment above and below the dam. The releases from Lake Powell, primarily made through the powerplant, were colder and clearer than the original river flow. Also, the daily fluctuation was more dramatic and more frequent. The shift in the river's flow in turn changed the species composition in the river and the habitat. A significant issue facing the EIS team was the presence of both native and non-native species downstream from the dam.[220]

In addition to studying and reporting on the general impact of the project, the EIS also offered recommendations for changing the dam's operation. The EIS offered several alternative operating procedures[221] designed to lessen or mitigate the impact upon the native fish species, including the pikeminnow,[222] humpback chub, and razorback sucker, and benefit the downstream environment. At the same time, the releases from the dam needed to conform to the other beneficial uses of the project, including recreation and power generation, and adhere to the requirements of the Colorado River Compact.

Several of the suggested alternatives included beach/habitat-building flows,

Removing the rotor from one of the generators in Shasta Powerplant. Central Valley Project: Shasta/Trinity River Division. Date: 1997. Photographer: Tom Papowich.

Operation of jet-flow valves, Glen Canyon Dam. Colorado River Storage Project: Glen Canyon Unit, Arizona, Utah. Date: March 1996. Photographer: Unknown

or scheduled high water releases of short duration, designed to rebuild high-elevation sandbars, deposit nutrients, restore backwater channels, and provide some of the dynamics of a natural system. These high water releases were designed to simulate the natural flood patterns of the river. The EIS further stated that a test flow would be made to analyze the predictions of the modelers.

Secretary Bruce Babbitt in front of the jet tubes during a test flow at Glen Canyon Dam. Colorado River Storage Project: Glen Canyon Unit. Date: March 1996. Photographer: Tom Smart. Photo courtesy of *Deseret News.*

In a nationally publicized event, the outlet works were opened, increasing flows through the powerplant at Glen Canyon Dam to release water in a simulated natural flood for seven days in March of 1996.[223] At the time of the test, Secretary of the Interior Bruce Babbitt, among others, declared the experiment a success; however, only long-term monitoring and study will determine the final success or failure of the project. This ongoing process is called an adaptive management program.

Concurrent with the EIS process, in 1994, the U.S. Fish and Wildlife Service (FWS) issued a biological opinion recommending that Reclamation investigate the feasibility of installing temperature control modifications[224] on the existing intake structure. Such modifications allow for water releases of varying temperatures by drawing water from various levels of the reservoir; existing intake structures draw exclusively from the cold depths of the reservoir. In response to the FWS suggestion, Reclamation began preparing an environmental assessment (EA) on the proposed temperature control modifications and the potential impacts. Reclamation completed a draft EA in January of 1999, including a peer review process. The final decision on the fifteen-million dollar proposed modifications had not yet been made when this book went to press.[225]

POWER MANAGEMENT LABORATORY

Beginning in 1993, President Clinton's Administration attempted to redefine government, making it more consumer friendly and cost effective. To oversee this effort, Clinton formed the National Performance Review (NPR) under the direction of Vice President Al Gore. At the same time, the impending issue of deregulation of the electric utility industry brought to light questions about the marketing, delivery, and value of electric energy in a market-driven environment. In an attempt to address both issues, Interior authorized Reclamation to take part in the NPR and form the Power Management Laboratory (Laboratory). The purpose of the Laboratory was to identify how Reclamation could reinvent and better manage its hydropower program to better meet customer needs.

The Power Management Laboratory itself consisted of several small work groups or teams, each responsible for identification and assessment of a specific area for potential improvement within the power program.[226] A management team – comprised of representatives from throughout Reclamation, as well as customers, partners, the environmental community, stakeholders, and an independent analyst from Argonne National Laboratories – oversaw the activities of the smaller work groups. The teams used benchmarks, surveys, and work group analysis to examine the power program, then made recommendations for improvement.

In 1997, the National Performance Review recognized Reclamation's continuing efforts to improve the efficiency of its power facilities by presenting the Power Management Laboratory with the prestigious Hammer Award, given to teams of Federal employees that had shown significant progress in reinventing government. The Laboratory was honored for its unprecedented, comprehensive review of Reclamation's hydropower program as detailed in the publication *Future Generations: A New Era of Power, Performance, and Progress*. Though the Administration has changed, Reclamation continues to serve western energy needs as efficiently and economically as possible.

BENCHMARKING

One of the most significant undertakings of the Power Management Laboratory was an eighteen-month benchmarking[227] review of Reclamation's hydropower program, beginning in 1995. By comparing Reclamation's program to others in the industry, this effort painted a comprehensive picture of the power program's standing within the hydropower industry. Benchmarks also provided a baseline for evaluation of future improvements and modifications to the power program.

Reclamation's Power Program defined specific areas for study, or benchmarks, based on recommendations and industry standards. The benchmarks were then narrowed down to seven specific areas for focus and potential improvement: how much we charge our customers; what percentage of our customer charges cover our operating costs; how much it costs us to generate electricity; the number of staff required to generate electricity; how often our generators are available for use; unscheduled maintenance; and scheduled maintenance. These seven benchmarks provided the framework within which Reclamation studied its own performance and its standing within the industry.

Teams compared Reclamation powerplants with both Reclamation and non-Reclamation plants of similar capacity and generation type to determine each plant's current and potential capabilities. The Power Management Laboratory used the results of the benchmarking study to recommend methods, approaches, and actions for reaching and surpassing industry standards. Future results and changes could then be measured against the original benchmarks. Through a continual benchmarking process, Reclamation continued to evaluate and improve the performance of its power facilities. Overall,

Working on Unit No. 2 in Black Canyon Powerplant after rewinding. The generators, built in 1921, each deliver 4,000 kW of power. Boise Project, Idaho. Date: January 10, 1994. Photographer: David Walsh.

Overhauling Unit No. 3 at Yellowtail Powerplant. Two cranes linked by a special beam provide the lift capacity needed to pull the rotor out of the generator. Overhaul of the unit required remanufacturing of parts to compensate for normal wear. Pick-Sloan Missouri Basin Program: Yellowtail Unit, Montana. Date: 2001. Photographer: Jerry Leggate.

Reclamation's hydropower program performed favorably in comparison to the industry at large. Production costs and system reliability remained well within industry standards.[228]

DEREGULATION

Until very recently, the electric power market has been fairly all-inclusive. A utility was responsible for the generation, transmission, and distribution of power to its customers that fell within state-designated service areas. State laws governed or regulated investor-owned utility operations and permitted utilities to do whatever was necessary to meet growing customer demand within their service area, including constructing additional generation, transmission, and distribution facilities.

State governments, in the form of public utility commissions or public service commissions, regulated shareholder-owned utility transactions on the retail level. These state commissions, comprised of elected or appointed officials, made sure all utility expenditures were reasonable. This in turn allowed the utility to recoup these expenditures from its customers through its rates. Commissions also regulated an investor-owned utility's rate of return, ensured that utilities responded to customers' needs and service requests, and sustained adequate and reliable service. Federal, state, and local government-owned utilities and member-owned electric cooperatives do not fall under these commissions' regulations.

On the wholesale level, the Federal Energy Regulatory Commission regulated all interstate transactions. As on the retail level, FERC's regulation did not apply to government-owned utilities or electric cooperatives.

With the passage of the Public Utilities Regulatory Policy Act of 1978, the traditional method of doing business began to change. The tightly controlled process of generating, transmitting, and distributing electricity started to disintegrate. No longer could utilities just build new powerplants to meet growing needs.

After PURPA went into effect, FERC mandated that utilities buy power from "qualifying facilities"[229] at a price equal to the costs of building a new powerplant or purchasing energy from another source. FERC set the initial standards for pricing of electricity generated at qualifying facilities.[230] Many state utility commissions took the process one step further by requiring utilities to purchase energy regardless of need. PURPA essentially created a special relationship between certain power generators and utilities, not to mention a new level of control by the Federal government.

In the 1990s, the nature of the electric utility market began to change once again. After passage of the Energy Policy Act of 1992, a new entity known as power marketers started competing with utilities for customers at both the wholesale and retail level.[231] This new group of utility participants expanded the traditional vertically integrated monopolies into a more diverse utility structure with many more members.

The transmission of energy across a utility's system also falls under FERC's jurisdiction. FERC now ensures guaranteed open access to interstate high-voltage wires of shareholder-owned utilities for all suppliers and sets fair and reasonable rates for transmission services. In 1996, FERC issued Order Nos. 888 and 889, which require shareholder-owned utilities to open their transmission lines to all suppliers in the wholesale market. This means any wholesale electrical energy supplier can use transmission lines owned by shareholder-owned utilities at the same price and terms those utilities charge themselves to transmit power.

Order No. 889 also required utilities to separate their generation or transmission and distribution capabilities over a number of years, which further divided the control a utility had over the industry. At the same time, this divestiture of resources was supposed to encourage smaller companies to become part of and invest in the industry. Order Nos. 888 and 889 essentially opened the electric utility market to broad-based wholesale competition.

The final effort to dissolve the monopoly-like structure of the utility industry occurred late in 1999, when FERC issued Order No. 2000. This order encouraged transmission-owning utilities to turn control of their transmission systems over to regional transmission organizations by the end of the year 2001.[232] In many cases, these new organizations are taking the form of regional transmission organizations, Independent System Operators, or Independent Scheduling Agents. These organizations will help maintain the integrity and stability of the power grid, at the same time taking control of the transmission network from the utilities.

At the time this manuscript was published, Reclamation did not know the ultimate results of deregulation on its general business practices. The direct impacts will fall upon Western and BPA, which market and sell Federal power in a competitive marketplace. What steps these organizations will need to take to adapt to the new marketplace, continuing downward pressure on cost, and the resulting demands on generation, remain to be seen. The relationship among transmission, generation, and distribution means that any necessary restructuring or change in the way the power marketing administrations do business will likely affect Reclamation also.

YEAR 2000 ISSUES

Beginning as early as 1998, the power community at large, and Reclamation

Installing the turbine head cover in Unit No. 3, Yellowtail Powerplant. Pick-Sloan Missouri Basin Program: Yellowtail Unit, Montana. Date: 2001. Photographer: Jerry Leggate.

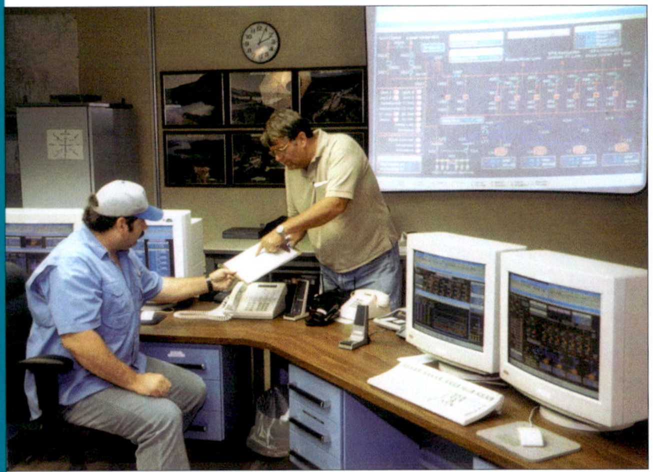

SCADA system operators at Black Canyon Dam. Boise Project, Idaho. Date: 2001. Photographer: David Walsh.

specifically, began to address the potential for year 2000 problems, commonly referred to as Y2K, or the Y2K bug.[233] Reclamation formed an embedded microchip (EMC)[234] team to prevent failure of systems on all potential Y2K dates, including January 1, 2000. Staff from Reclamation's Technical Service Center, supported by an advisory group, formed the basis of the team. The team was responsible for finding and reprogramming or replacing all the embedded microchips throughout Reclamation to ensure a smooth transition to the new year. This included verifying that Supervisory Control and Data Acquisition (SCADA) systems[235] were Y2K ready. Because of the importance of SCADA systems, a separate team was formed that provided guidance and technical expertise in testing and certifying those systems.

Two years of effort went into determining Reclamation's general Y2K readiness. The technical staff developed special strategies to deal with the unique configurations of Reclamation's systems. All work done by the team was subject to the nationally accepted standards for Y2K and February 29th (leap day). These standards consisted of awareness, assessment, testing, renovation, validation, and contingency planning.

All told, the team inventoried 1,688 embedded microchip devices, of which 883 of these were "mission critical" devices requiring testing, certification, validation, and contingency plans.

Additionally, independent verification and validation was conducted for all the mission critical chips.

Reclamation inventoried and assessed all fourteen of its SCADA systems. Ten of these systems received a mission critical rating, requiring extensive testing and in some cases renovation. Testing often took place at both the component and system level. As with the EMC effort, independent verification and validation was performed. The team also checked an additional thirteen SCADA systems built by Reclamation but transferred to other entities.

In addition to inventorying and testing, Reclamation, in coordination with BPA and Western, crafted a Power Operations Contingency Plan for every Reclamation powerplant. In April and September of 1999, Reclamation successfully participated in NERC-sponsored power system drills. All activities took place in conjunction with the industry and the Power Marketing Administrations.

In the end, all of Reclamation's preparations paid off, as no serious problems were encountered during the January 1, 2000, or February 29, 2000, date changes.

Footnotes for Chapter 11: Reclamation and the Energy Crisis

[213] Rose, xv; Casazza, 38-44.

[214] A North Central Power Study identified potential renewable energy sources, including potential locations for peaking hydroplants and pump storage sites in Colorado, Montana, North and South Dakota, Utah, and Wyoming. Some of the sites investigated gained additional notice at the start of the energy crisis. However, none of the projects identified were ever constructed. *North Central Power Study, Report of Phase 1*, North Central Power Study Coordinating Committee, October 1971.

[215] <http://www.brakenhale.berks.sch.uk/wfs/energy/renew.htm> accessed September 2, 1998; *Hydropower: Water at Work*, 7.

[216] "Power Production at Federal Dams Could Be Increased by Modernizing Turbines and Generators."

[217] A rewind is defined as the act of putting new copper-insulated wire in the armature windings of a generator. Reclamation purchased many of its older generators with a continuous overload capacity of fifteen percent above rated output (nameplate capacity). Generally, Reclamation purchases a new winding with a base rating equal to the 115 percent machine level, as well as an appropriate allowable temperature rise consistent with the insulation class of the new winding. Reclamation then replaces an old winding with a new one. Furthermore, because the mechanical characteristics of the generator are not confirmed to support higher loads, the machine is limited to operation at the 115 percent level. Additionally, Reclamation examines ratings of the bus, transformer, and other equipment.

[218] <http://www.usbr.gov/power/data/uprate/uprate.htm, 1-8> accessed August 25, 1998.

[219] Cooperating entities were the Bureau of Indian Affairs, National Park Service U.S. Fish and Wildlife Service, Western Area Power Administration, Arizona Game and Fish Department, Hopi Tribe, Hualapai Tribe, Navajo Nation, San Juan Southern Paiute Tribe, Southern Paiute Consortium, and Zuni Pueblo. Representatives from most of these cooperators and a private consulting firm formed the actual interdisciplinary EIS team responsible for the study.

[220] The non-native species downstream from the dam are part of the blue ribbon trout fishery which developed in part due to the cold water released from the dam. Part of the challenge to the EIS team was balancing the needs of the trout fishery with the needs of the native species.

[221] Formally entitled an Adaptive Management Program, the EIS recommended that Reclamation form a Federal advisory committee to oversee the future management of the dam and assess the downstream effects of the structure. The charter for this group was signed on January 15, 1997. <http://www.uc.usbr.gov/amp/backgrnd.html> accessed March 9, 2000.

[222] The pikeminnow was formerly known as the squawfish.

[223] Prior to construction of the dam, annual maximum daily flows of water greater than 80,000 cfs were common. The test flow was a special release of up to 45,000 cfs for a maximum duration of seven days. The release was preceded and followed by four days of low, steady, 8,000 cfs flows.

To simulate the natural flood patterns, the flows were gradually increased until they reached their peak of 45,000 cfs and then gradually decreased.

[224] Reclamation previously installed temperature control devices on several of its dams, including Flaming Gorge, Shasta, Hungry Horse, and Jordanelle. Installed in 1978, the controls at Flaming Gorge resulted in a positive impact on the river system below the dam. As a result of the warmer water, trout growth rates immediately below the dam increased significantly; native fish downstream near the Yampa River also showed improvement. <http://www.uc.usbr.gov/amp/wr/gctc_feature.html> accessed March 9, 2000.

[225] For additional information see United States Department of the Interior, Bureau of Reclamation, *Operation of Glen Canyon Dam: Final Environmental Impact Statement* (Government Printing Office, March 1995); United States Department of the Interior, Bureau of Reclamation, *Glen Canyon Dam, Beach/Habitat-Building Flow: Final Environmental Assessment and Finding of No Significant Impact* (Salt Lake City: Upper Colorado Regional Office, February 1996).

[226] The management team defined six major components of the power program: cost containment; environmental compatibility; financial accountability; human resources; operations and maintenance; and power management. A specific work group was assigned a specific component for comprehensive study.

[227] Reclamation defined benchmarking as a formal process of comparison, in specifically defined areas, between Reclamation and other similar utilities to determine its standing within the industry at large.

[228] *Future Generations*, A-1-A-6.

[229] PURPA created a special class of power producers known as Qualifying Facilities, which fell into two categories: Small Power Producers and Cogenerators. Small Power Producers were limited in size to 80 mW and had to use a renewable energy source to generate at least seventy-five percent of its total power. Cogenerators had to use at least five percent of the steam generated by a fuel source, including fossil fuels, for two distinct purposes: electricity generation and a separate commercial or industrial use. The separate use was usually heating.

Qualifying facilities were also exempt from cost of service regulations under state and Federal law and from the financial and organizational requirements of PUHCA. See earlier chapters for a complete discussion of PUHCA.

[230] By establishing standards for pricing of Qualifying Facilities' electricity, FERC ensured that utilities and electricity consumers did not pay higher prices for power under the new system. The standards established resulted in rates which reflected a utility's "avoided costs" – the costs (including the construction of new powerplants) a utility avoided by purchasing power from a qualifying facility.

FERC took things one step further, however, when it set the avoided cost rate. Believing that fuel prices were going to increase, FERC set the avoided cost rate above the actual avoided costs at that time and far above current market value. Since most Qualifying Facilities' contracts were written for thirty years, the price inequities remain.

[231] Wholesale power is sold between utilities or bought from other power suppliers for resale to the end user. Retail power is sold directly to the end user, such as a home or business.

232 <http://www.eei.org/issues/comp_reg/key_facts/section2.pdf>; <http://www.eei.org/issues/comp_reg/regulation.pdf>; <http://www.eei.org/issues/comp_reg/PURPA.pdf>; <http://www.eei.org/issues/comp_reg/competition.pdf> all accessed March 9, 2000.

233 When originally designing and programming computers, most programmers and designers only used the last two digits of the year to denote the date. As the transition to the year 2000 approached, the issue of how systems would recognize the 00 date became apparent. The fact that the year 2000 was a leap year also presented potential problems. Many within the computer industry believed that most systems would recognize the 00 date as 1900 instead of 2000. The then inaccurate date held the potential to corrupt existing and new data. Corrupted data can produce error messages which can in turn force systems to shut down either manually or automatically. International attention focused on the potential for failure of essential systems, including utilities, military defense, and communications. To combat the problem, companies and governments around the world mobilized to deal with the issue.

234 Computers often contain embedded microchips responsible for operation of the system. These chips sometimes contained date codes, even if the chip itself did not require a date for operation. Those date codes, in turn, might conflict with the operation of the system.

235 SCADA systems are the computer systems that allow for the remote control of powerplants.

"Generator - Shasta Dam." Central Valley Project: Shasta/Trinity Unit, California. Date: Unknown. Artist: Roland Peterson.

CHAPTER TWELVE

Originally a by-product of irrigation, flood control, and river regulation, hydropower development became increasingly more influential in determining the function, benefits, impacts, and scale of Reclamation's projects. Hydroelectric energy production at Reclamation's storage dams helped finance larger irrigation projects and encouraged comprehensive river basin development. Electricity from Reclamation's projects helped carry the country through times like the Great Depression, World War II, the Cold War, and the Korean Conflict. These occasions of economic and political crisis encouraged the Federal government's role in the development of power. Meanwhile, Reclamation hydroelectric energy production addressed the growing trends toward domestic and industrial electrification.

Through its various multipurpose construction projects, Reclamation helped to shape the West's future. One of its most impressive features, Hoover Dam, remains one of the most widely known water and hydroelectric projects in the world. Every year, more than one million people travel from around the world to tour the structure. Many more unrecorded visitors choose not to take the tour, content simply to view the dam from atop. In 1955, the American Society of Civil Engineers selected Hoover as one of its seven civil engineering wonders of the United States, acknowledging the enduring importance of the project to the American West. The society also included Grand Coulee Dam and the Columbia Basin Project and the Colorado River Aqueduct of the Metropolitan Water District of Southern California, for which Reclamation constructed Parker Dam on the Colorado River, on its list of seven structures. In 1999, a major construction exposition recognized Hoover as number five on the Top Ten Construction Achievements of the Twentieth Century. In southern Nevada, Hoover consistently receives annual recognition as the non-gaming attraction most visited by the public. In 2000, both Hoover and Grand Coulee were recognized by the American Public Works Association as Top Ten Public Works Projects of the Twentieth Century. In 2001, the American Society of Civil Engineers selected Hoover Dam as one of the civil engineering projects of the millennium.

Reclamation contributed to the modernization and urbanization of the West by providing inexpensive electrical energy. Often, Reclamation power drew new industries into an area, offering alternatives to a previously entirely agrarian lifestyle. Even as Reclamation continues to move from a large-scale construction bureau to a water management bureau, hydropower generation remains an integral part of its overall mission.

Reclamation continues to look at new technologies and management techniques to improve operation and maintenance and powerplant reliability, while improving efficiency and minimizing environmental risks. Emerging high-voltage generation technology has been investigated and

The four generators inside the Palisades Powerplant. Total output of the four units is 118,750 kW. Palisades Project, Idaho, Wyoming. Date: March 4, 1982. Photographer: Glade Walker.

may prove beneficial in powerplant rehabilitation – expensive and troublesome equipment may possibly be eliminated. Also, Reclamation's powerplant Life Extension program will allow for better management decisions for plant modernization through the use of risk management and economic evaluation coupled with modern methods of equipment diagnostics.

The Bureau of Reclamation remains a leader in the hydroelectric industry. Its hydropower plants provide system stability and reliability to the western utility system. In 2000, it was the tenth largest utility in the country, providing surplus energy for sale by Western and BPA. All told, an excess of thirteen-billion dollars was returned to the U.S. Treasury for repayment of funds previously advanced for construction, after payment of annual operation, maintenance, and repair costs. As Reclamation recognized its Centennial in 2002, it continued to be a leader in the electrical utility industry.

Transmission lines leading from Hoover Dam strech across the Mojave Desert. Date: 1941. Photographer: Ansel Adams.

Old and new generators in the Shoshone Powerplant. The powerplant went into service in 1922. Beginning in 1982, the plant was rehabilitated and the new generator (at back) was added. Shoshone Project: Buffalo Bill Dam Rehabilitation, Wyoming. Date: August 28, 2001. Photographer: Jerry Leggate.

Light show at Grand Coulee Dam. Columbia Basin Project, Washington. Date: September 17, 1989. Photographer: Unknown.

Beck, Bill. "Arc Lighting and the Birth of Public Power." *Public Power* (July-August 1998).

Bennett, Newcomb. "A New Era Of Power Transmission." *Reclamation Era* 51 (August 1965): 72-75.

Berber, Michele Stenehjem. *On the Home Front: The Cold War Legacy of the Hanford Nuclear Site*. Lincoln: University of Nebraska Press, 1997.

Billington, David P., Donald C. Jackson, and Martin Melosi. *Federal Multipurpose Dams: The U.S. Army Corps of Engineers and the Bureau of Reclamation in the Era of Big Dams*. Draft Manuscript: December 22, 2000.

Brigham, Jay L. *Politics and Power: The Fight for Electricity in the West, 1902-1932.* Lawrence, Kansas: University of Kansas Press, 1998.

Brown, D. Clayton. *Electricity for Rural America: The Fight for the REA*. Westport, Connecticut: Greenwood Press, 1980.

Cannon, Brian Q. "'We Are Now Entering a New Era': Federal Reclamation and the Fact Finding Commission of 1923-1924." *Pacific Historical Review* 66 (May 1997): 185-211.

Casazza, John A. *The Development of Electric Power Transmission*. New York: IEEE, 1993.

Cheney, Margaret. *Tesla: Man Out of Time*. New York: Barnes and Noble Books, 1993.

Cohen, Lizabeth. *Making a New Deal: Industrial Workers in Chicago, 1919-1939.* Chicago: Cambridge University Press, 1990.

Coleman, Charles M. *P.G. and E. of California: The Centennial Story of Pacific Gas and Electric Company, 1852-1952.* New York: McGraw-Hill Book Company, 1952.

Cooke, Morris L. "Giant Power: Large Scale Electrical Development as a Social Factor." *The Annals of the American Academy of Political and Social Science* 118 (March 1925).

Cooper, Gail. *Air-Conditioning America: Engineers and the Controlled Environment, 1900-1960*. Baltimore: Johns Hopkins University Press, 1998.

Cowan, Ruth Schwartz. *A Social History of a American Technology*. New York: Oxford University Press, 1996.

_____. *More Work for Mother: The Ironies of Household Technology from the Open Hearth to the Microwave*. New York: Basic Books, 1983.

Craeger, William P., and Joel D. Justin. *Hydroelectric Handbook*. New York: John Wiley and Sons, Inc., 2nd edition, 1950.

Curran, Christine. "A Historic Context for the Transmission of Hydroelectricity by the Bonneville Power Administration, 1939-1945." M.A. Thesis, University of Oregon, 1998.

DeGraaf, Leonard. "Corporate Liberalism and Electric Power System Planning in the 1920s." *Business History Review* 64 (Spring 1990): 1-31.

Dominy, Floyd E. "A New Power Giant Materializes on the West Coast." *Reclamation Era* 51 (August 1965): 63.

Downing, P.M. "Foundations for Transmission Line Towers and Tower Erection." *Proceedings of the American Institute of Electrical Engineers* (June 20, 1915): 1227-1233.

Farmer, Jared. *Glen Canyon Dammed: Inventing Lake Powell and the Canyon Country*. Tucson: University of Arizona Press, 1999.

Ficken, Robert E. "Grand Coulee and Hanford: The Atomic Bomb and the Development of the Columbia River." In *The Atomic West*. Bruce Hevly and John M. Findlay, eds. Seattle: University of Washington Press, 1998.

_____. *Rufus Woods: The Columbia River and the Building of Modern Washington*. Pullman, Washington: Washington State University Press, 1995.

Fradkin, Philip L. *A River No More: The Colorado River and the West*. New York: Alfred A. Knopf, 1981.

Frederick, Kenneth D., and Roger A. Sedjo, eds. *America's Renewable Energy Resources: Historic Trends and Current Challenges*. Washington D.C.: Resources for the Future, 1991.

Friedel, Robert and Paul Israel. *Edison's Electric Light: Biography of an Invention*. New Brunswick, New Jersey: Rutgers University Press, 1986.

Funigiello, Philip. *Toward a National Power Policy: The New Deal and the Electric Utility Industry, 1933-1941*. Pittsburgh, Pennsylvania: University of Pittsburgh Press, 1973.

_____. "Kilowatts for Defense: The New Deal and the Coming of the Second World War." *The Journal of American History* 56:3 (December 1969): 604-620.

Gaylord, James M. "Power and Pumping System of the Salt River Project, Arizona." Phoenix: Salt River Project Archives, January 1, 1914.

Glaser, Leah S. *The EMA Transmission Line*. Historic American Engineering Record, National Park Service, No. AZ-6-B, 1996.

______. "The Navajo Indian Irrigation Project." Denver: Bureau of Reclamation History Program, 1998.

Golźe, Alfred R. *Reclamation in the United States*. Caldwell, Idaho: Caxton Printers, Ltd., 1961.

Goodland, Robert. "Environmental Sustainability in the Hydro Industry: Disaggregating the Debate." *Large Dams: Learning From the Past Looking at the Future.* Tony Dorcey, ed. Gland, Switzerland: The World Conservation Union, April 1997.

Gottleib, Robert, and Margaret FitzSimmons. *Thirst for Growth: Water Agencies as Hidden Government in California*. Tucson: University of Arizona Press, 1991.

Harvey, Mark W.T. *A Symbol of Wilderness: Echo Park and the American Conservation Movement*. Albuquerque: University of New Mexico Press, 1994.

Hay, Duncan. *Hydroelectric Development in the United States 1880-1940.* Washington, D.C.: Edison Electric Institute, 1991.

Hays, Samuel P. *Conservation and the Gospel of Efficiency: The Progressive Conservation Movement, 1890-1920*. Cambridge: Harvard University Press, 1959.

Hevly, Bruce, and John M. Findlay, eds. *The Atomic West*. Seattle: University of Washington Press, 1998.

Hirsh, Richard F. *Technology and Transformation in the American Electric Utility Industry*. Cambridge, Massachusetts: MIT Press, 1989.

Hoffman, Abraham. *Vision or Villainy: Origins of the Owens Valley – Los Angeles Water Controversy*. College Station: Texas A&M University Press, 1981.

Hoy, Suellen M., Michael C. Robinson, and Ellis L. Armstrong. *History of Public Works in the United States, 1776-1976*. Chicago, Illinois: American Public Works Association, 1976.

Hughes, Thomas P. *American Genesis: A Century of Invention and Technological Enthusiasm, 1870-1970*. New York: Viking University Press, 1989.

______. *Networks of Power: Electrification in Western Society, 1880-1930.* Baltimore, Maryland: The Johns Hopkins University Press, 1983, 1988.

Hundley, Norris, jr. "Clio Nods: *Arizona v. California* and the Boulder Canyon Act – A Reassessment." *Western Historical Quarterly* 3 (January 1972): 17-51.

_____. *The Great Thirst: Californians and Water, 1770s-1990s*. Berkeley: University of California Press, 1992.

_____. *Water and the West: The Colorado River Compact and the Politics of Water in the American West*. Berkeley: University of California Press, 1975.

_____. "The Politics of Reclamation: California, the Federal Government, and the Origins of the Boulder Canyon Act – A Second Look." *California Historical Quarterly* 52 (Winter 1973): 292-325.

Hunter, Louis C. *A History of Industrial Power in the United States, vol. 1, 2 & 3.* Charlottesville: University of Virginia Press, 1979.

International Union for Conservation of Nature and Natural Resources/The World Bank. *Large Dams: Learning from the Past, Looking at the Future*. Tony Dorcey, ed. Gaithersburg, Maryland: Reproductions, Inc., 1997.

Introcaso, David. *Mormon Flat Dam*. Historic American Engineering Record, National Park Service, No. AZ-14, 1989.

Jackson, Donald C. *Building the Ultimate Dam: John S. Eastwood and the Control of Water in the West*. Berkeley and Los Angeles: University of California Press, 1975.

James, George Wharton. *Reclaiming the Arid West: The Story of the United States Reclamation Service*. New York: Dodd, Mead, and Company, 1917.

Kasson, John F. *Civilizing the Machine: Technology and Republican Values in America, 1776-1900*. New York: Grossman, 1976.

Kathka, David Arlin. "The Bureau of Reclamation in the Truman Administration: Personnel, Politics, and Policy." Ph.D. Dissertation, University of Missouri, 1976.

Keller, Charles. *The Power Situation During the War*. Washington, D.C.: Government Printing Office, 1921.

Klotz, Louis H. "Water Power and its Promises." Durham, New Hampshire: University of New Hampshire Center for Industrial and Institutional Development, February 1980.

Komarek de Luna, Phyllis. *Public Versus Private Power During the Truman Administration: A Study of Fair Deal Liberalism*. Modern American History Series. New York and Bern: Peter Lang, 1997.

Kyle, John H. *The Building of TVA: An Illustrated History*. Baton Rouge: Louisiana State University Press, 1958.

Lawson, Michael L. *Dammed Indians: The Pick-Sloan Plan and the Missouri River Sioux, 1944-1980.* Norman: University of Oklahoma Press, 1982.

Layton, Edwin T., Jr. "Scientific Technology, 1845-1900: The Hydraulic Turbine and the Origins of American Industrial Research." *Technology and Culture* 20 (January 1979): 78-89.

Leuchtenberg, William E. *Franklin D. Roosevelt and the New Deal: 1932-1940*. New York: Harper and Row Publishers, 1963.

Lewis, Christine. "A History of Irrigation on the Tempe Area." M.A. Thesis, Arizona State University, 1963.

Lindseth, Emil V. "Researching the Intertie." *Reclamation Era* 51 (August 1965): 69.

Linenberger, Toni Rae. "A Dam for All Seasons: Hollywood, the Bureau of Reclamation, and the Construction of Parker Dam." M.S. Report, Utah State University, 1998.

_____. "The Navajo Unit: Colorado River Storage Project." Denver: Bureau of Reclamation History Program, 1998.

Lowitt, Richard. *The New Deal and the West*. Bloomington: Indiana University Press, 1984.

Mast, Gerald. *A Short History of the Movies*. Fourth Edition. New York: Macmillan Publishing Company, 1986.

McElvaine, Robert S. *The Great Depression: America, 1929-1941*. New York: Times Books, 1993, second edition.

Melosi, Martin V. *Thomas A. Edison and the Modernization of America*. Glenview, Illinois: Scott Foresman, 1990.

_____. *Coping with Abundance: Energy and the Environment in Industrial America*. New York: Alfred A. Knopf, 1985.

Mermel, Thaddeus. "D-C Developments in Other Countries." *Reclamation Era* 51 (August 1965): 86-89.

Mighetto, Lisa. "Salmon, Science, and Politics: Writing History for the U.S. Army Corps of Engineers." *The Public Historian* 17 (Fall 1995): 22-24.

Murray, Vernon M. "Grand Coulee and Bonneville Power in the National War Effort." *The Journal of Land and Public Utility Economics* 18 (February-November 1942): 134-139.

Myers, William A. *Iron Men and Copper Wires: A Centennial History of the Southern California Edison Company*. Glendale, California: Trans-Anglo Books, 1983.

Nash, Gerald D. "Science in the Wartime West." In *The American West Transformed: The Impact of the Second World War*. Lincoln: University of Nebraska Press, 1985.

Nash, Roderick. *The Nervous Generation, 1917-1930*. Chicago: Elephant Paperbacks, Ivan R. Dee Publishers, 1970.

Noble, David F. *America by Design: Science, Technology, and the Rise of Corporate Capitalism*. New York: Alfred A. Knopf, 1977.

North American Electric Reliability Council. *NERC 1998 Annual Report*.

Norwood, Gus. *Columbia River Power for the People: A History of Policies of the Bonneville Power Administration*. Portland, Oregon: Department of Energy, Bonneville Power Administration.

Nye, David E. *American Technological Sublime.* Cambridge, Massachusetts: The MIT Press, 1994.

______. *Consuming Power: A Social History of American Energies*. Cambridge, Massachusetts: The MIT Press, 1998.

______. *Electrifying America: Social Meanings of a New Technology, 1880-1940.* Cambridge, Massachusetts: The MIT Press, 1990.

______. "Electrifying the West." In *The American West Reconsidered*. Rob Kroes, ed. Amsterdam: Free University Press, 1989.

O'Neill, Dan. "Alaska and the Firecracker Boys: The Story of Project Chariot." In *The Atomic West*. Bruce Hevly and John M. Findlay, eds. Seattle: University of Washington Press, 1998.

Par Environmental Services, Inc. *Cultural Resources Investigation for the American River Bridge Crossing Project, City of Folsom, Sacramento County, California*. Sacramento, CA: Par Environmental Services, Inc., 1991.

Pendergrass, Bonnie Baack. *Public Power, Politics, and Technology in the Eisenhower and Kennedy Years: The Hanford Dual-Purpose Reactor Controversy, 1956-1962*. New York: Arno Press, 1979.

Pfaff, Christine. *San Carlos Irrigation Project*. Historic American Engineering Record, National Park Service, No. AZ-50.

_____. *Theodore Roosevelt Dam, Power Plant*. Historic American Engineering Record, National Park Service, No. AZ-6-A, 1996.

Pisani, Donald J. *From Family to Agribusiness: The Irrigation Crusade in California and the West, 1850-1931*. Berkeley: University of California Press, 1984.

_____. *To Reclaim a Divided West: Water, Law, and Public Policy, 1848-1902*. Albuquerque: University of New Mexico Press, 1992.

Pitzer, Paul C. *Grand Coulee: Harnessing a Dream*. Pullman, Washington: Washington State University Press, 1994.

_____. "The Mystique of Grand Coulee Dam and the Reality of the Columbia Basin Project." Paper, Western History Association. Tacoma, Washington: October 14, 1989.

Pursell , Carroll. *The Machine in America: A Social History of Technology*. Baltimore, Maryland: The Johns Hopkins University Press, 1995.

_____. "Government and Technology in the Great Depression." *Technology and Culture* 20 (January 1979): 162-174.

Reich, Leonard S. *The Making of American Industrial Research: Science and Business at GE and Bell, 1876-1926*. New York: Cambridge University Press, 1985.

Reisner, Marc. *Cadillac Desert: The American West and Its Disappearing Water*. New York: Viking and Penguin Books, 1986.

Reynolds, Terry S. "Hydro Before the Turbine: The Rise and Fall of the Water Wheel." *Hydro Review* 11 (December 1992): 52.

Ridley, Scott. *Profile of Power: A History of the People and Events That Have Shaped and Continue to Shape America's Most Critical Industry*. Washington, D.C.: American Public Power Association, 1996.

Riebsame, William E., Hannah Gosnell, and David Theobald, eds. *Atlas of the New West: Portrait of a Changing Region*. New York: WW Norton and Company, 1997.

Righter, Robert W. *Wind Energy in America: A History*. Norman: University of Oklahoma Press, 1996.

Robinson, Michael. *Water for the West: The Bureau of Reclamation, 1902-1977*. Chicago, Illinois: Public Works Historical Society, 1979.

Roosevelt, Theodore. An Autobiography. New York: Macmillan, 1913.

Rose, Mark H. *Cities of Light and Heat: Domesticating Gas and Electricity in Urban America*. University Park, Pennsylvania: Pennsylvania State University Press, 1995.

Rosenberg, Nathan. *Technology and American Economic Growth.* White Plains, New York: M.E. Sharpe, 1972.

Rudolph, Richard, and Scott Ridley. *Power Struggle: The Hundred-Year War Over Electricity*. New York: Harper and Row, 1986.

Rushmore, David B. and Eric A. Lof. *Hydro-Electric Power Stations*. New York: John Wiley and Sons, 1923.

Rydell, Robert W. *All the World's a Fair: Visions of Empire at American International Expositions, 1876-1916*. Chicago, Illinois: University of Chicago Press, 1984.

_____. *World of Fairs: The Century of Progress Expositions*. Chicago, Illinois: University of Chicago Press, 1993.

Schatz, Thomas. *The Genius of the System: Hollywood Filmmaking in the Studio Era*. New York: Pantheon Books, 1988.

Scholes, D.R. "Fundamental Considerations Governing the Design of Transmission-line Structures." *Proceedings of the American Institute of Electrical Engineers* (June 29, 1908): 1011-1018.

Sheridan, Thomas. *Arizona: A History*. Tucson: The University of Arizona Press, 1995.

Simon, Charnan. *Hollywood at War: The Motion Picture Industry and World War II*. New York: Franklin Watts, 1995.

Simonds, Wm. Joe. *The Columbia Basin Project*. Denver: Bureau of Reclamation History Program, 1998.

Spence, Clark C. "Early Uses of Electricity in American Agriculture." *Technology and Culture* 3 (Spring 1962): 143.

Springer, Vera. *Power and the Pacific Northwest: A History of the Bonneville Power Administration*. United States Department of the Interior, 1976.

Stacy, Susan M. *Legacy of Light: A History of Idaho Power Company*. Boise, Idaho: Idaho Power Company, 1991.

Stevens, Joseph E. *Hoover Dam: An American Adventure*. Norman: University of Oklahoma Press, 1988.

Swanson, Mark T. *Santa Ana River Hydroelectric System*. Historic American Engineering Record, National Park Service, No. CA-130.

Thorson, John E. *River of Promise, River of Peril: The Politics of Managing the Missouri River*. Lawrence, Kansas: University Press of Kansas, 1994.

Tobey, Ronald. *Technology as Freedom: The New Deal and the Electrical Modernization of the American Home*. Berkeley: University of California Press, 1995.

Tollefson, Gene. *BPA and the Struggle for Power at Cost*. Portland, Oregon: Bonneville Power Administration, 1987.

United States Army Corps of Engineers. *The History of the U.S. Army Corps of Engineers*. Alexandria, Virginia, 1998.

______. *Big Dam Era: A Legislative and Institutional History of the Pick-Sloan Missouri Basin Program*. By John R. Ferrell. Omaha, Nebraska: Missouri River Division, 1993.

United States Department of the Interior, Harold L. Ickes, Secretary. "The War Program of the Department of the Interior." Washington, D.C.: The National Archives Library, c. 1942.

United States Department of the Interior, Bureau of Reclamation. *Federal Reclamation and Related Laws Annotated*. Volume I of IV, through 1942. Richard K. Pelz, ed. Washington, D.C.: U.S. Government Printing Office, 1972.

______. *Flaming Gorge Dam and Powerplant: Technical Record of Design and Construction*. Denver: Government Printing Office, 1968.

______. *Future Generations: A New Era of Power, Performance, and Progress.* 1996.

______. *Glen Canyon Dam, Beach/Habitat-Building Flow: Final Environmental Assessment and Finding of No Significant Impact*. Salt Lake City: Upper Colorado Regional Office, February 1996.

______. *Hoover Dam*. Washington D.C.: Government Printing Office, 1985.

______. *Hydropower: Water at Work*. Washington, D.C.: Government Printing Office, c. 1985.

______. *Operation of Glen Canyon Dam: Final Environmental Impact Statement*. Government Printing Office, March 1995.

______. *Reclamation's Energy Initiatives: Taking a Leadership Role in Supporting the National Energy Strategy*. Draft. October 1990.

_____. *Reclamation's Role in Hydropower*. Briefing paper for Fernando Xavier Fereirra. Denver: Bureau of Reclamation, June 4, 1991.

_____. *Repayment of Reclamation Projects*. Washington, D.C.: Government Printing Office, 1972.

United States Department of the Interior, Water and Power Resources Service. *Project Data*. Denver: Government Printing Office, 1981.

Waltrip, John Richard. *Public Power During the Truman Administration*. New York: Arno Press, 1979.

Warne, William E. *The Bureau of Reclamation*. Boulder, Colorado: Westview Press, 1973.

Wellock, Thomas Raymond. *Critical Masses: Opposition to Nuclear Power in California, 1958-1978*. Madison: University of Wisconsin Press, 1998.

White, Richard. *The Organic Machine*. New York: Hill and Wang, 1995.

Williams, James. *Energy and the Making of Modern California*. Akron: University of Ohio Press, 1997.

SELECTED ARTICLES

Annual Reports of the Secretary of the Interior. Washington D.C.: Government Printing Office, 1927-1963.

Proceedings for the American Institute of Electrical Engineers.

Project Histories, RG 115, Rocky Mountain Region: National Archives and Records Administration, Denver, Colorado.

The Public Historian.

Reclamation Era.

WEBSITES CONSULTED

www.brakenhale.berks.sch.uk/wfs/energy/renew

www.earthday.org

www.eei.org/issues/comp_reg/key_facts/section2

www.eei.org/issues/comp_reg/regulation

www.eei.org/issues/comp_reg/PURPA

www.eei.org/issues/comp_reg/competition

www.glencanyon.net/navajo

www.nellis.af.mil/history/default

www.technocracy.org

www.townhall.com/pff/pucha/part2

www.uc.usbr.gov/amp/backgrnd

www.usbr.gov/main/programs/native-am

www.usbr.gov/power/data/uprate/uprate

www.usbr.gov/power/edu/history

www.usoe.k12.ut.us/curr/science/core/8thgrd/integrated/Potentialkinetic

www.wapa.gov/western/factsw

"Dam." Date: Unknown. Artist: Allan D'Arcangelo.

Reclamation's fifty-eight powerplants have made a significant contribution to the seventeen western states, both on an individual level and as a whole. In addition, Reclamation built several plants for other entities, which also made significant contributions to local economies.

Reclamation powerplants annually generate over forty-two billion kWh of hydroelectric energy, enough to meet the annual residential needs of fourteen million people or the energy equivalent of over eighty million barrels of crude oil. Revenues from the sale of Reclamation energy are now nearly one billion dollars annually. Since 1909 power revenues have contributed over ten billion dollars in project repayment to the Federal Treasury.

Each of Reclamation's powerplants has its own unique story. While it was not possible to highlight each powerplant's story within the body of the text, the following appendix contains a brief description and short history of each facility.

For additional information on Reclamation's Power Program, please contact:

Deborah Linke, Manager
Power Resources Office
Bureau of Reclamation
PO Box 25007 (D5400)
Denver, CO 80225
303-445-2921
dlinke@do.usbr.gov
power@do.usbr.gov

Or visit the website at:

www.usbr.gov/power

For additional information on Reclamation's History Program, please visit the website at:

www.usbr.gov/history

Alcova Powerplant

Great Plains Region
Kendrick Project

Location: Alcova Powerplant is located in Natrona County, Wyoming, on the North Platte River, about ten miles downstream from Pathfinder Dam. The powerplant is on the right bank of the river, opposite the toe of the dam.

Authorization: The President approved the Kendrick Project on August 30, 1935. The Alcova Powerplant was authorized for construction on August 22, 1950, under the provisions of section 9(a) of the Reclamation Project Act of 1939. Originally known as Casper-Alcova, the project was renamed Kendrick in 1937.

Description: Water from Alcova Dam is released for other irrigation rights downstream through the Alcova Powerplant or over a controlled spillway.

History: Alcova Powerplant was authorized and built after Alcova Dam was completed in 1938. Construction of Alcova Powerplant was completed in 1955.

Anderson Ranch Powerplant

Pacific Northwest Region
Boise Project

Location: Anderson Ranch Powerplant is located in Elmore County, Idaho, on the South Fork of the Boise River, about twenty miles northeast of Mountain Home, Idaho.

Authorization: Construction of the original Boise Project was authorized on March 27, 1905. The Secretary of the Interior authorized Anderson Ranch Dam and Reservoir on August 12, 1940, under the Reclamation Project Act of 1939 (53 Stat. 1187).

Description: Anderson Ranch Dam and Powerplant is a multipurpose structure that provides benefits in irrigation, power, and flood and silt control. The plant supplies power for irrigation needs in the Boise, Owyhee, and Minidoka Projects as part of Reclamation's Southern Idaho Power System. Surplus power is delivered to the Bonneville Power Administration for marketing and distribution to regional industries and municipalities.

History: The first right to divert water from the Boise River for irrigation purposes was granted in 1864. The water irrigated the town site of Boise and supplied Fort Boise. Agricultural activity in the Boise and Payette Valleys started in the early 1880s when settlers began filing for desert lands under private irrigation enterprises. By 1900, about 148,000 acres had been placed under irrigation. It quickly became evident that storage and distribution systems were needed for a dependable water source to serve the lands at high elevations. In response to petitions by local irrigators, the Boise Project was initiated by the Reclamation Service shortly after the passage of the first Reclamation Act of 1902.

Big Thompson Powerplant

Great Plains Region
Colorado-Big Thompson Project

Location: The Big Thompson Powerplant is on the Big Thompson River, about nine miles west of Loveland, Colorado, and just downstream from the river crossing of the Charles Hanson Feeder Canal. The Big Thompson Powerplant is located in Larimer County, Colorado.

Authorization: First construction funds were provided by the Interior Department Appropriation Act of August 9, 1937 (50 Stat. 595). The President approved the Secretary's finding of feasibility on December 21, 1937.

Description: A seasonal powerplant used to generate peaking power during the summer months. Water from either Flatiron Reservoir or the Big Thompson River runs from the Feeder Canal down 180 feet to the single unit plant before it is returned to the river.

History: Began operation in 1959. The Big Thompson Plant generates approximately fifteen million kWh a summer.

Black Canyon Powerplant

Pacific Northwest Region
Boise Project

Location: Black Canyon Powerplant is located in Gem County, Idaho, on the Payette River, about five miles northeast of Emmett, Idaho.

Authorization: The Secretary of the Interior authorized construction of the original Boise Project on March 27, 1905, and the Black Canyon Dam on June 26, 1922, under provision of the Reclamation Act of June 17, 1902 (32 Stat. 388).

Description: The plant supplies power for irrigation needs in the Boise, Owyhee, and Minidoka Projects as part of Reclamation's Southern Idaho Power System. Surplus power is delivered to the Bonneville Power Administration for marketing and distribution to regional industries and municipalities.

History: Agricultural activity in the Boise and Payette Valleys started in the early 1880s when settlers began filing for desert lands under private irrigation enterprises. By 1900, about 148,000 acres had been placed under irrigation. Black Canyon Dam is part of the Payette Division of the Boise Project.

Blue Mesa Powerplant

Upper Colorado Region
Colorado River Storage Project

Location: The Blue Mesa Powerplant is located on the Gunnison River in Gunnison County, about thirty miles west of Gunnison, Colorado, and about one and a half miles below Sapinero, Colorado.

Authorization: Congress authorized construction of four storage units of the Colorado River Storage Project and eleven participating projects on April 11, 1956, under Public Law 485, 84th Congress, 70 Stat. 105. Additional projects have been added since the original legislation was adopted.

Description: Generation of hydroelectric power for the upper basin and adjacent areas.

History: Power generation was initiated in September of 1967 at the Blue Mesa Powerplant. Generators were uprated from 30,000 kW to 43,200 kW in 1988.

Boise River Diversion Powerplant

Pacific Northwest Region
Boise Project

Location: Boise River Diversion Powerplant is located in Ada County, Idaho, on the Boise River, about seven miles southeast of Boise, Idaho.

Authorization: The original Boise Project was authorized on March 27, 1905. The Arrowrock Dam was authorized on January 6, 1911.

Description: When operating, the plant supplies power for irrigation needs in the Boise, Owyhee, and Minidoka Projects as part of Reclamation's Southern Idaho Power System. Surplus power is delivered to the Bonneville Power Administration for marketing and distribution to regional industries and municipalities.

History: Power production commenced in 1912, providing electic power for the construction of Arrowrock Dam, with surplus being utilized in the growing Boise Valley. The plant is currently in the process of a complete renovation, through which the existing equipment will be refurbished, updated, and automated, while retaining the plant's historic quality.

Boysen Powerplant

Great Plains Region
Pick-Sloan Missouri Basin Program

Location: Boysen Powerplant is located in Fremont County, Wyoming, approximately twenty miles south of Thermopolis, Wyoming, on the Wind River.

Authorization: The Congress authorized the Boysen Unit of the Pick-Sloan Missouri Basin Program under the Flood Control Act of December 22, 1944, (Public Law 534), which approved the general comprehensive plan set forth in Senate Document 191, as revised and coordinated by Senate Document 247, 78th Congress, 2nd Session.

Description: Power generated at the Boysen Powerplant is fed into the Western Division, Pick-Sloan Missouri Basin Program transmission facilities for use within that division.

History: Construction of the powerplant was authorized by the Flood Control Act of December 22, 1944. Construction began in 1947 and was completed in 1952.

Buffalo Bill Powerplant

Great Plains Region
Shoshone Project

Location: Buffalo Bill Powerplant is located in Park County, Wyoming, approximately four miles southwest of Cody, Wyoming, on the Shoshone River.

Authorization: The Secretary of the Interior authorized the Shoshone Project on February 10, 1904, under authority of the Reclamation Act of June 17, 1902, and authorized Heart Mountain power development on June 19, 1945, under the provisions of the Reclamation Project Act of 1939.

Description: Waters of the Shoshone River are stored in Buffalo Bill Reservoir for later release for irrigation and power generation. Power is developed at the Shoshone, Heart Mountain, Buffalo Bill, and Spirit Mountain Powerplants. The system is interconnected with the West Division of the Pick-Sloan Missouri Basin Program.

History: In 1992, construction was completed on the dam modification (which raised the dam twenty-five feet), the new powerplant, and the visitors' center.

Canyon Ferry Powerplant

Great Plains Region
Pick-Sloan Missouri Basin Program

Location: Canyon Ferry Powerplant is located on the mainstem of the Missouri River about fifty-eight miles north of its headwaters. Helena, the capital of Montana, is seventeen miles southwest of the damsite.

Authorization: The Canyon Ferry Unit of the Pick-Sloan Missouri Basin Program was authorized by the Flood Control Act of December 22, 1944 (Public Law 534), which approved the general comprehensive plan set forth in Senate Document 191, as revised and coordinated by Senate Document 247, 78th Congress, 2nd Session.

Description: Canyon Ferry Powerplant, with an installed capacity of 50,000 kW, supplies expanding power needs for commercial use in a wide surrounding area by interconnection with Montana Power Company facilities.

History: Canyon Ferry Powerplant was constructed to provide regulation of runoff for low-cost power and to permit increased diversions in the Upper Missouri River Basin.

Chandler Powerplant

Pacific Northwest Region
Yakima Project

Location: Chandler Powerplant is located in Benton County, Washington, about ten miles northeast of Prosser, Washington. It is on the Chandler power channel next to the Yakima River.

Authorization: The Congress authorized the Yakima Project, Kennewick Division, on June 12, 1948, under Public Law 629, 80th Congress (62 Stat. 382).

Description: The powerplant has not historically supplied project power for irrigation pumping. Electric pumps to replace the existing hydraulically powered pumps at the powerplant are planned for installation in another location.

History: The powerplant was constructed in 1956, as part of the Yakima Project.

Crystal Powerplant

Upper Colorado Region
Colorado River Storage Project

Location: The Crystal Powerplant is located on the Gunnison River in Montrose County, Colorado, six miles downstream of Morrow Point Powerplant and approximately twenty miles east of Montrose, Colorado.

Authorization: The Congress authorized construction of four storage units in the Colorado River Storage Project and eleven participating projects on April 11, 1956, under Public Law 485, 84th Congress, 70 Stat. 105. Additional projects have been added since the original legislation was adopted.

Description: Generation of hydroelectric power for use in the project.

History: Transformer KRA was replaced in fiscal year 1996.

Lower Colorado Region
Parker-Davis Project

Location: Davis Dam spans the Colorado River in Pyramid Canyon, sixty-seven miles downstream from Hoover Dam and eighty-eight miles upstream from Parker Dam. Davis Dam is in Mohave County, Arizona, and Clark County, Nevada, on the Nevada-Arizona state line, thirty-two miles west of Kingman, Arizona, and one mile north of Laughlin, Nevada.

Authorization: The Parker-Davis Project was formed by consolidating the Parker Dam Power Project and the Davis Dam Project under the terms of the Act of May 28, 1954 (68 Stat. 143). The Secretary of the Interior authorized Davis Dam on April 26, 1941, under provisions of the Reclamation Project Act of 1939 (53 Stat. 1187).

Description: The electrical integration and interconnection of Davis, Hoover, and Parker Powerplants provide maximum generation of power with efficient use of water resources. The highly developed agricultural base and the complex industrialization of the Pacific Southwest benefit greatly from Colorado River hydroelectric energy.

History: The project was originally authorized as the Bullshead Project by the Department of the Interior's Appropriation Act of 1942. The name was changed to Davis Dam in honor of one of Reclamation's directors, Arthur Powell Davis.

Deer Creek Powerplant

Upper Colorado Region
Provo River Project

Location: The Deer Creek Powerplant is located on the Provo River in Wasatch County, Utah, about sixteen miles northeast of Provo, Utah.

Authorization: Construction of the project was initiated under the provisions of the National Industrial Recovery Act of June 16, 1933. The President approved the Provo River Project on November 16, 1935, under the terms of subsection B of section 4 of the Act of December 5, 1942 (43 Stat. 701). The President approved the Salt Lake Aqueduct on October 24, 1938. The Secretary of the Interior authorized the Deer Creek Powerplant on August 20, 1951, under the Reclamation Project Act of 1939.

Description: The powerplant provides replacement energy for energy lost at other non-government plants through the storage and diversion of water for irrigation during the winter storage season. The powerplant provides revenues during the summer irrigation season for operation, maintenance, and replacement costs and for irrigation project repayment purposes. When power is not used for replacement purposes, it is marketed to two preference power entities.

History: The project is essentially an irrigation project with production of power limited to the capacity of the plant during the irrigation season. Maintenance is normally performed during the winter season when water flows are low. Since 1958, the powerplant has been operated and maintained by a contract with the Provo River Water Users Association.

Elephant Butte Powerplant

Upper Colorado Region
Rio Grande Project

Location: Elephant Butte Powerplant is located on the Rio Grande River in Sierra County, New Mexico, four miles east of Truth or Consequences, New Mexico, and 125 miles north of El Paso, Texas.

Authorization: The Secretary of the Interior authorized the Rio Grande Project on December 2, 1905, under the provision of the Reclamation Act, and funds were allocated to initiate construction of the first diversion unit. The Reclamation Act was extended to the entire state of Texas on June 12, 1906, following a partial extension for Engle (Elephant Butte) Dam in 1905. Congress authorized the construction of Elephant Butte Dam on February 25, 1905; and on May 4, 1907, one million dollars of nonreimbursable funds were appropriated as the State Department's share for allocation of 60,000 acre-feet of water annually to Mexico by treaty. Additional project works authorized under congressional action include Caballo Dam, a combined flood control and power regulating structure, and the Elephant Butte power development.

Description: Generates hydroelectric power for the project.

History: A system consisting of 490 miles of 115 kV transmission line and eleven substations totaling 81,750 kilovolt-amperes was developed and operated by the Rio Grande Project until 1977. It was subsequently sold to a private electric company.

Estes Powerplant

Great Plains Region
Colorado-Big Thompson Project

Location: Estes Powerplant is located in Larimer County, Colorado, in Estes Park, Colorado.

Authorization: The President approved the Secretary of the Interior's finding of feasibility on December 21, 1937.

Description: Estes Powerplant contains three generating units served by three seventy-eight inch-diameter penstocks about three quarters of a mile long. Water diverted from Colorado's west slope flows through a thirteen mile-long tunnel under the Continental Divide, then through a smaller siphon and tunnel to Marys Lake, before falling through the penstocks to the Estes Powerplant. Like all the powerplants on the east slope of the Colorado-Big Thompson project, Estes is an off-stream facility. Once power has been generated, water exits the plant to fill Lake Estes.

History: Reclamation built a camp for its workers in Estes Park in the late 1940s when construction on the powerplant began. Construction workers left town when the plant was completed in 1950, but some of the original camp houses are still there.

Flaming Gorge Powerplant

Upper Colorado Region
Colorado River Storage Project

Location: The Flaming Gorge Powerplant is located on the Green River in Daggett County, Utah, twenty-six miles downstream from the Utah-Wyoming border.

Authorization: Congress authorized construction of four storage units in the Colorado River Storage Project and eleven participating projects on April 11, 1956, under Public Law 485, 84th Congress, 70 Stat. 105. Additional projects have been added since the original legislation was adopted.

Description: As the Flaming Gorge Dam is on a major tributary to the Colorado River, it was constructed for long-term storage of water and controlled releases to meet downstream commitments. The powerplant's purpose is to use this stored water for the economical generation of electricity and provide the revenue for repayment of the project.

History: Construction was authorized on April 11, 1956, and construction began in 1958, with road building, a diversion tunnel, and other preparatory work. The first unit went online September 27, 1963, and the third and final unit in February of 1964. The Flaming Gorge Reservoir first filled in August of 1974 at elevation 6,040 with 3,749,000 acre-feet of live storage. The plant was automated in 1973 and 1974, and was then controlled by the Colorado River Storage Project Dispatch Center located in Montrose, Colorado. Selective withdrawal structures were installed on the turbine penstock intakes during the winters of 1977 and 1978 to provide temperature control of the water released down the Green River to improve fishery habitat. A temperature of fifty-five degrees Fahrenheit can be maintained from mid-June into October. On August 11, 1977, the Unit No. 2 turbine seized up when the lower seal ring failed. This incident led to the replacement of seal rings on all three units. The newly designed and installed seal rings also failed and were again replaced on each unit. Generator uprates began in August of 1990 and were completed in April of 1992. The generators were uprated from 36,000 kW to 50,650 kW.

Flatiron Power and Pumping Plant

Great Plains Region
Colorado-Big Thompson Project

Location: Flatiron Power and Pumping Plant is located in Larimer County, Colorado, near Carter Lake, ten miles west of Loveland, Colorado.

Authorization: The President approved the Secretary of the Interior's finding of feasibility on December 21, 1937.

Description: Two penstocks carry water down a 1055-foot drop to Flatiron Powerplant. The plant contains two main power units and a reversible 13,000-horsepower pump-turbine unit that lifts water south from Flatiron to Carter Lake through a 1.4-mile-long pressure tunnel. During peak power load demands, water can be released from Carter Lake back into Flatiron through the reversible unit.

History: Began operation in 1954. Water discharged through Flatiron is stored for irrigation deliveries in Carter, Flatiron, and Horsetooth Reservoirs.

Folsom Powerplant

Mid-Pacific Region
Central Valley Project

Location: Folsom Powerplant is located on the American River in Sacramento County, California, about twenty miles northeast of Sacramento, California.

Authorization: Funds for construction of the initial features of the Central Valley Project were provided by the Emergency Relief Appropriation Act of 1935 (49 Stat. 115). The Secretary of the Interior authorized the project, and the President approved it on December 2, 1935.

Description: Folsom is a peaking powerplant which is dedicated first to meeting the requirements of the project facilities. The remaining energy is marketed to various preference customers in northern California. This plant also provides power for the pumping plant, which supplies the local domestic water supply.

History: Folsom Dam was constructed by the Corps of Engineers and upon completion was transferred to the Bureau of Reclamation for coordinated operation as an integral part of the Central Valley Project. Construction of the dam began in October of 1948 and was completed in May of 1956. The units were uprated in 1972.

Fontenelle Powerplant

Upper Colorado Region
Seedskadee Project

Location: Fontenelle Powerplant is located on the Green River in Lincoln County, Wyoming, twenty-four miles southeast of LaBarge, Wyoming.

Authorization: The Seedskadee Project is one of the initial group of participating projects authorized with the Colorado River Storage Project on April 11, 1956 (70 Stat. 105).

Description: Fontenelle Powerplant provides maximum generation of power from regulated river flows depending upon water availability. Flows vary from 350 cfs to 1,550 cfs. Generation varies from one megawatt at low seasonal head to eleven megawatts at peak runoff periods.

History: From 1986-1989, the powerplant was shut down while the reservoir was being drained to install a diaphragm cutoff wall for the length of the dam. In June of 1989, the unit began generating again at new low-head capabilities. This was in part due to the addition of additional riprap on the upstream face of the dam.

Fremont Canyon Powerplant

Great Plains Region
Pick-Sloan Missouri Basin Program

Location: Fremont Canyon Powerplant is located in Natrona County, Wyoming, approximately three miles downstream from Pathfinder Dam, on the left bank of the North Platte River Canyon.

Authorization: The Glendo Unit of the Pick-Sloan Missouri Basin Program was authorized by the Flood Control Act of December 22, 1944 (Public Law 534), which approved the general plan set forth in Senate Document 191, as revised and coordinated by Senate Document 247, 78th Congress, 2nd Session. The project was reauthorized by Public Law 503, 83rd Congress, on July 16, 1954.

Description: The powerplant generates power during releases of stored water from Pathfinder Reservoir on the North Platte Project.

History: Construction of Fremont Canyon Powerplant and power conduit began in 1956 and was completed in 1961. The plant had an original installed capacity of 48,000 kW, and was uprated between 1986 and 1990, through the installation of new generator windings and turbines, to its current installed rating of 66,800 kW.

Glen Canyon Powerplant

Upper Colorado Region
Colorado River Storage Project

Location: The Glen Canyon Powerplant is located on the Colorado River in Coconino County, Arizona, near the Utah border. It is fifteen miles upstream from Lees Ferry.

Authorization: Congress authorized construction of four storage units in the Colorado River Storage Project and eleven participating projects on April 11, 1956, under Public Law 485, 84th Congress, 70 Stat. 105. Additional projects have been added since the original legislation was adopted.

Description: Generates hydroelectric power for the project.

History: Generator uprating was completed between 1984 and 1987, and Unit No. 8 was uprated in 1997. Five generators are rated at 165,000 kW each, and three generators are rated at 157,000 kW each.

Glendo Powerplant

Great Plains Region
Pick-Sloan Missouri Basin Program

Location: Glendo Powerplant is located in Platte County, Wyoming, approximately four and a half miles southeast of Glendo, Wyoming, on the North Platte River.

Authorization: The Glendo Unit of the Pick-Sloan Missouri Basin Program was authorized by the Flood Control Act of December 22, 1944 (Public Law 534), which approved the general plan set forth in Senate Document 191, as revised and coordinated by Senate Document 247, 78th Congress, 2nd Session. The project was reauthorized by Public Law 503, 83rd Congress, on July 16, 1954.

Description: Addition of Glendo Unit power generation facilities increased available power in the North Platte River Basin by about 500 million kilowatt hours annually. This increase comes primarily from the Glendo and Fremont Canyon Powerplants. Some of the gain, however, was due to the conversion of the Alcova Powerplant from seasonal to year-round operation, made possible by the regulation afforded by Glendo Reservoir.

History: The plant had an original installed capacity of 24,000 kW and was uprated between 1980-1984 to its present installed capacity of 38,000 kW by replacing turbines and windings.

Grand Coulee Powerplant

Pacific Northwest Region
Columbia Basin Project

Location: Grand Coulee Powerplant is located in Grant and Okanogan Counties, Washington, on the Columbia River, about twenty-eight miles northeast of Coulee City, Washington.

Authorization: Funds were first allocated for the Columbia Basin Project and Grand Coulee Dam under the National Industrial Recovery Act of June 16, 1933. The project was specifically authorized for construction by the Rivers and Harbors Act approved August 30, 1935. The Columbia Basin Project Act of March 10, 1943 (57 Stat. 14), reauthorized the project, bringing it under the provisions of the Reclamation Project Act of 1939. Construction of the Third Powerplant was authorized by Public Law 89-448 (80 Stat. 200), dated June 14, 1966, as amended by Public Law 89-561 (80 Stat. 714), dated September 7, 1966.

Description: Grand Coulee Dam is a key feature of Reclamation's Columbia Basin Project in central Washington. A multipurpose project, it provides flood control, irrigation, hydropower production, recreation, stream flows, and fish and wildlife benefits. Facilities at the dam include three powerplants, a pump generating plant, and three switchyards. Franklin D. Roosevelt Lake, behind the dam, is 151 miles long with over 5,000,000 acre-feet of active storage. An average of 2,500,000 acre-feet of water is pumped annually for irrigation. Water is pumped in the Columbia Basin to irrigate approximately 670,000 acres.

History: Construction of the original project started in 1933 and was completed in 1942. The first power was generated in 1941, and the last of the original eighteen units (Right and Left Powerplants) began production in 1950. Pumping for irrigation commenced in 1952, with six pumps installed. Six pump generators were installed from the mid-1970s (P/G-7 and P/G-8) to the early 1980s (P/G-9 through P/G-12). Construction of the Third Powerplant and Forebay Dam commenced in 1967 with the first unit (G-19) commissioned in 1975 and the last (G-24) in 1980. The eighteen original generators in the Right and Left Powerplants have had the stator windings replaced, increasing the rating from the original 108,000 kW to 125,000 kW. The last three generators in the Third Powerplant (G-22, G-23, and G-24) had their stators and cores replaced, increasing their rating to 805 mW. Four banks (twelve) of the original transformers in the Left Powerplant have been replaced, with the ratings increased to 129,000 kilovolt amperes.

Great Plains Region
Colorado-Big Thompson Project

Location: Green Mountain Powerplant is located in Summit County, Colorado, thirteen miles southeast of the town of Kremmling, Colorado. The plant is on the Blue River, a tributary of the Colorado.

Authorization: The President approved the Secretary of the Interior's finding of feasibility on December 21, 1937.

Description: Green Mountain Dam provides replacement storage for water diverted by the Colorado-Big Thompson Project from Colorado's western slope to the eastern slope. It is the project's only "on-stream" reservoir. The powerplant has two units with a total installed generating capacity of 21,600 kilowatts.

History: Green Mountain was the first facility to be construction on the C-BT project. Construction began in 1938 and concluded in 1943. Power was also first generated in 1943.

Green Springs Powerplant

Pacific Northwest Region
Rogue River Basin Project

Location: Green Springs Powerplant is located in Jackson County, Oregon, about ten miles east of Ashland, Oregon.

Authorization: Congress authorized the Talent Division of the Rogue River Basin Project in Public Law 606, 83rd Congress, 2nd Session, approved August 20, 1954 (68 Stat. 752).

Description: Green Springs Powerplant was constructed to take advantage of the head drop of the transbasin diversion to produce power for commercial sales to assist repayment of the division costs.

History: The Green Springs Powerplant was placed in service in 1960 and has helped meet expanding power demands in southern Oregon and northern California. Revenues have contributed significantly to the repayment of construction costs of the Talent Division. The generator was rewound and the switchyard rehabilitated in 1993.

Guernsey Powerplant

Great Plains Region
North Platte Project

Location: Guernsey Powerplant is located in Platte County, Wyoming, approximately two miles upstream of Guernsey, Wyoming, on the North Platte River, about 180 miles below Alcova Dam and twenty-five miles below Glendo Dam.

Authorization: The Secretary of the Interior authorized the project on March 14, 1903. The President approved the Guernsey Dam and Powerplant on April 30, 1925, as a part of the North Platte Project, which was originally called the Sweetwater Project.

Description: The Guernsey Dam controls river flow. Water released from Pathfinder Reservoir can be stored at this dam and released to fit varying irrigation demands. Water is released through Guernsey Powerplant and outlet works.

History: Construction of Guernsey Powerplant began in 1925 and was completed in 1928. The original installed capacity of the plant was 4,800 kW. The plant was uprated between 1992 and 1994 through replacement of the generator windings to its current installed capacity of 6,400 kW.

Heart Mountain Powerplant

Great Plains Region
Shoshone Project

Location: Heart Mountain Powerplant is located in Park County, Wyoming, approximately four miles southwest of Cody, Wyoming, on the Shoshone River.

Authorization: The Secretary of the Interior authorized the project on February 10, 1904, under authority of the Reclamation Act of June 17, 1902. The Secretary authorized Heart Mountain power development on June 19, 1945, under the provisions of the Reclamation Project Act of 1939.

Description: Power produced at the Heart Mountain Powerplant is fed into a grid system, which serves an area extending into three states.

History: Construction of the powerplant, which was originally built to be a temporary plant, was completed in 1947. The plant's generator was rewound in 1992.

Hoover Powerplant

Lower Colorado Region
Boulder Canyon Project

Location: Hoover Powerplant is located in the Black Canyon of the Colorado River, about thirty-six miles from Las Vegas, Nevada, on the Arizona-Nevada state line.

Authorization: The project was authorized by the Boulder Canyon Project Act of December 21, 1928 (45 Stat. 1057), subject to the terms of the Colorado River Compact. The Boulder Canyon Project Adjustment Act (54 Stat. 774), dated July 19, 1940, provided for certain changes to the original plan. The Hoover Powerplant Act of 1984 (98 Stat. 1333) provided for the uprating of the generators and the construction of new visitor facilities.

Description: The primary purpose of the electrical powerplant is to generate revenue sufficient to repay the project construction monies advanced by the U.S. Treasury and to annually fund ongoing operation, maintenance, and replacement expenses. The capacity and operational flexibility of the plant also significantly contribute to the stability and reliability of the California, Arizona, Nevada, Utah, and New Mexico electrical grid. The electrical integration and interconnection of Davis, Hoover, and Parker Powerplants provide maximum generation of power with efficient use of water resources. The highly developed agricultural base and the complex industrialization of the Pacific southwest benefit greatly from Colorado River hydroelectric energy.

History: The plant was authorized as a Boulder Canyon Project feature (the act also provided for construction of the All-American Canal and nearby Boulder City). The first commercial unit to be placed in service was N-2 on October 26, 1936. The last unit to be commercially commissioned was N-8 on December 1, 1961. There were eleven commercially available units by 1939. The plant was operated by Southern California Edison and the Los Angeles Department of Water and Power under the supervision of the Bureau of Reclamation until 1987, the termination date of the original fifty year electric service contracts. Reclamation assumed control of operation and maintenance in 1987.

The Boulder Canyon Project Adjustment Act of 1940 authorized the Secretary of the Interior to promulgate and implement charges for electrical energy generated at Hoover Powerplant. This act further specified how the revenues resulting from Hoover energy sales were to be applied. The Hoover Powerplant Act of 1984 authorized the uprating of the seventeen main generating units, construction of additional visitor facilities and parking structures, and the Hoover Dam bypass bridge. It also changed the funding source from Congressional Appropriations to an Available Receipts Fund (a quasi-revolving fund). The uprating of the units was accomplished through non-Federal funds advanced by eleven power customers.

The dam was originally designated as Hoover Dam by instruction of the Secretary of the Interior, dated September 17, 1930. The dam was designated Boulder Dam by order of the Secretary, dated May 8, 1933. The name Hoover Dam was restored by the Act of April 30, 1947 (61 Stat. 56).

Hungry Horse Powerplant

Pacific Northwest Region
Hungry Horse Project

Location: Hungry Horse Powerplant is located in Flathead County, Montana, on the South Fork of the Flathead River, about nine miles southeast of Columbia Falls, Montana.

Authorization: Congress authorized construction of Hungry Horse Dam under Public Law 329, 78th Congress, 2nd Session, approved June 5, 1944 (58 Stat. 270).

Description: Hungry Horse Powerplant creates power benefits that extend from the Continental Divide westward to the Pacific Ocean. The principal power benefit from the project arises from its ability to store water through the spring flood season for later release.

History: The generating units were uprated and overhauled from 1990-1993. The capacity of each unit increased from 71,250 kW to 107,000 kW. Conversion work for remote operation began in 1994. The switchyard was rebuilt in 1995, and a penstock selective withdrawal system was also installed in 1995.

Judge F. Carr Powerplant

Mid-Pacific Region
Central Valley Project

Location: Judge Francis Carr Powerplant is located on Clear Creek in Shasta County, California, at the outlet of Clear Creek Tunnel, on the northwestern extremity of Whiskeytown Lake.

Authorization: Funds for construction of the initial features of the Central Valley Project were provided by the Emergency Relief Appropriation Act of 1935 (49 Stat. 115). The Secretary of the Interior authorized the project, and the President approved it on December 2, 1935.

Description: Judge Francis Carr Powerplant is a peaking plant which is dedicated first to meeting the energy requirements of the project facilities. The remaining energy is marketed to various preference customers in northern California.

History: Judge Francis Carr was originally designated Clear Creek Powerplant. The units were uprated in 1984.

Keswick Powerplant

Mid-Pacific Region
Central Valley Project

Location: Keswick Powerplant is located on the Sacramento River in Shasta County, California, nine miles downstream from Shasta Dam, about four miles west of Redding, California.

Authorization: Funds for construction of the initial features of the Central Valley Project were provided by the Emergency Relief Appropriation Act of 1935 (49 Stat. 115). The Secretary of the Interior authorized the project, and the President approved it on December 2, 1935.

Description: Keswick Powerplant is a run-of-the-river plant which is dedicated first to meeting the energy requirements of the project facilities. The remaining energy is marketed to various preference customers in northern California.

History: The units were uprated in 1992.

Kortes Powerplant

Great Plains Region
Pick-Sloan Missouri Basin Program

Location: Kortes Powerplant is located in Carbon County, Wyoming, approximately two miles below Seminoe Dam on the North Platte River and about sixty miles southwest of Casper, Wyoming.

Authorization: Kortes powerplant, was found feasible by the Secretary of the Interior as a supplement to the Kendrick Project on November 26, 1941. It was authorized by the Flood Control Act of December 22, 1944 (Public Law 534), which approved the general plan set forth in Senate Document 191, as revised and coordinated by Senate Document 247, 78th Congress, 2nd Session.

Description: Because of the enormous increase in power demands in the area, and power sales commitments, an accelerated power program was developed. Generating equipment and machinery were constructed concurrently with Kortes Dam and powerplant.

History: Unit No. 2 was rewound in 1973. Units Nos. 1 and 3 were rewound in 1985.

Lewiston Powerplant

Mid-Pacific Region
Central Valley Project

Location: Lewiston Powerplant is located on the Trinity River, about seven miles downstream from Trinity Dam.

Authorization: Funds for construction of the initial features of the Central Valley Project were provided by the Emergency Relief Appropriation Act of 1935 (49 Stat. 115). The Secretary of the Interior authorized the project, and the President approved it on December 2, 1935. The Shasta and Trinity River Division of the Central Valley Project was authorized by Public Law 386, 84th Congress, 1st Session, approved August 12, 1955.

Description: Lewiston Powerplant is a run-of-the-river plant which provides station service to Trinity Powerplant. It also provides power to local fish hatcheries.

History: The powerplant originally served fish hatcheries and station service requirements for Trinity and Judge Francis Carr Powerplants. Administration of the contract was transferred to Western in 1977, and an interconnection contract with Pacific Gas and Electric was canceled in 1989. A new interconnection contract was signed in 1990.

Lower Molina Powerplant

Upper Colorado Region
Collbran Project

Location: The Lower Molina Powerplant is located in Mesa County, Colorado, on the south bank of Plateau Creek near Molina, Colorado.

Authorization: Congress authorized the Collbran Project on July 3, 1952, under Public Law 445, 82nd Congress, 2nd Session.

Description: Electrical energy is generated for use in west-central Colorado.

History: In fiscal year 1992, the unit control system was upgraded. A UPS system, static exciter, SCADA, and Sy/Max 400 PLC were installed in Lower Molina Powerplant. This reconfigured the plant from local operation to remote.

Marys Lake Powerplant

Great Plains Region
Colorado-Big Thompson Project

Location: Marys Lake Powerplant is located in Larimer County, Colorado, on the western shore of Marys Lake, two and a half miles southwest of Estes Park, Colorado.

Authorization: The Secretary's finding of feasibility was approved by the President on December 21, 1937.

Description: Water diverted from the west slope of Colorado enters Marys Lake Powerplant through a steel penstock 205 feet long. The small lake functions as both an afterbay for the Marys Lake plant and as a forebay for the Estes Powerplant.

History: Originally a small natural lake, Marys was enlarged by the construction of two dikes. Construction of the dikes and powerplant was completed in 1951.

McPhee Powerplant

Upper Colorado Region
Dolores Project

Location: McPhee Powerplant is located ten miles north of Dolores, Colorado, in Montezuma County, Colorado, on the Dolores River.

Authorization: Congress authorized the Dolores Project under the Colorado River Basin Act of September 30, 1968 (Public Law 90-537), as a participating project under the Colorado River Storage Project Act of April 11, 1956 (Public Law 84-485).

Description: McPhee Powerplant provides energy for project use.

History: The powerplant is operated and maintained under a contract with the Dolores Water Conservancy District. The powerplant was out of service from September of 1994 to January of 1997 due to a turbine problem.

Minidoka Powerplant

Pacific Northwest Region
Minidoka Project

Location: Minidoka Powerplant is located in Minidoka and Cassia Counties, Idaho, on the Snake River, about twenty-five miles northeast of Burley, Idaho.

Authorization: The Secretary of the Interior authorized the Minidoka Project on April 23, 1904. The Upper Snake River storage was authorized by a finding of feasibility by the Secretary and approved by the President on September 20, 1935.

Description: Natural flow of the Snake River and some of its tributaries, and water stored in the reservoirs at Jackson Lake, Grassy Lake, Island Park, American Falls, and Lake Walcott, are delivered at numerous diversion points to irrigation districts, American Falls Reservoir District, and Warren Act contractors. Much of the power developed on the project is used for pumping water to lands above the gravity canals and for pumping drainage water.

History: Minidoka Dam was originally designed and constructed without a powerplant and was completed in 1907. The powerplant and three pumping plants were added later (1909-1910), to allow water to be delivered to lands which could not be fed by gravity irrigation. The concrete buttress section of the dam forms the upstream structure of the original powerplant. The powerplant could only use a fraction of the water that passed through the dam, so Unit No. 6 was added in 1927, to allow commercial sales of power in excess of project needs. Unit No. 7 was added in 1942, to help feed the public's growing thirst for electric power. In September of 1995, the five original units were retired, and Unit Nos. 6 and 7 were rebuilt. Unit Nos. 8 and 9 were added in 1997 at Minidoka Dam, with the building of the Minidoka Replacement (now Allen E. Inman) Powerplant. The original powerplant building continues to house Unit Nos. 6 and 7.

Upper Colorado Region
Colorado River Storage Project

Location: The Morrow Point Powerplant is located on the Gunnison River in Montrose County, Colorado, twenty-two miles east of Montrose, Colorado.

Authorization: Congress authorized construction of four storage units in the Colorado River Storage Project and eleven participating projects on April 11, 1956, under Public Law 485, 84th Congress, 70 Stat. 105. Additional projects have been added since the original legislation was adopted.

Description: Generates hydroelectric power for the upper basin and adjacent areas.

History: The generators were uprated from 60,000 kW to 86,667 kW in 1992 and 1993. The three single-phase transformers were replaced in 1996, and the 230 kV cable was replaced in 1998.

Mt. Elbert Powerplant

Great Plains Region
Fryingpan-Arkansas Project

Location: Mt. Elbert Pumped-Storage Powerplant is located in Lake County, Colorado, on the north shore of Twin Lakes, twenty miles southwest of Leadville, Colorado. It is at the foot of Mt. Elbert, Colorado's highest peak.

Authorization: Congress authorized the Fryingpan-Arkansas Project under Public Law 87-590 (77 Stat. 393), signed by the President on August 16, 1962.

Description: The Mt. Elbert Pumped-Storage Powerplant is an all-concrete structure equivalent to a fourteen story building. Most of the structure is below ground. Water from the forebay above drops through two penstocks 445 feet to the powerplant where two turbine-generators develop 200,000 kilowatts of electrical power. Mt. Elbert's two units are also designed to operate as a 170,000-horsepower electric motor to drive the turbines in reverse and pump the same water back up to the forebay. This pumping mode is normally utilized during early morning hours when power demands are low. The pump-back storage principle is advantageous because the generating units can be started quickly and adjustments made rapidly to respond to varying daily and seasonal power demands.

History: Mt. Elbert is the only Federal powerplant on the Fryingpan-Arkansas Project. It began operations in 1981

New Melones Powerplant

Mid-Pacific Region
Central Valley Project

Location: The New Melones Powerplant is located in Tuolumne County, California, on the Stanislaus River.

Authorization: Funds for construction of the initial features of the Central Valley Project were provided by the Emergency Relief Appropriation Act of 1935 (49 Stat. 115). The Secretary of the Interior authorized the project, and the President approved it on December 2, 1935.

Description: New Melones Powerplant is a peaking plant. The power generated at this plant is dedicated first to meeting the requirements of the project facilities. The remaining energy is marketed to various preference customers in northern California.

History: In November of 1979, the New Melones Unit, East Side Division, was officially transferred to Reclamation from the Corps of Engineers by Public Law 87-874.

Nimbus Powerplant

Mid-Pacific Region
Central Valley Project

Location: The Nimbus Powerplant is located on the American River in Sacramento County, California, seven miles downstream from Folsom Dam. The powerplant is located on the right abutment of Nimbus Dam, on the north side of the river.

Authorization: Funds for construction of the initial features of the Central Valley Project were provided by the Emergency Relief Appropriation Act of 1935 (49 Stat. 115). The Secretary of the Interior authorized the project, and the President approved it on December 2, 1935.

Description: Nimbus Powerplant is a run-of-the-river plant and provides station service backup for Folsom Powerplant.

History: Normal operations are to maintain and regulate river releases.

O'Neill Pump-Generating Plant

Mid-Pacific Region
Central Valley Project

Location: O'Neill Pump-Generating Plant is located in Merced County, California, seventy miles from the Tracy Pumping Plant and twelve miles west of Los Banos, California. O'Neill Dam and Forebay are joint Federal-state facilities located on the San Luis Creek, two and a half miles downstream from San Luis Dam.

Authorization: Funds for construction of the initial features of the Central Valley Project were provided by the Emergency Relief Appropriation Act of 1935 (49 Stat. 115). The Secretary of the Interior authorized the project, and the President approved it on December 2, 1935. The San Luis Unit, West San Joaquin Division, was authorized as a part of the Central Valley Project on June 3, 1960 (Public Law 86-488).

Description: The O'Neill Pump-Generating Plant pumps Central Valley Project water for off-stream storage.

History: The pump-generating plant was turned over to Delta-Mendota Canal Authority for maintenance in 1994.

Palisades Powerplant

Pacific Northwest Region
Palisades Project

Location: Palisades Powerplant is located on the South Fork of the Snake River, about seven and a half miles southeast of Irwin, Idaho, and fifty-five miles southeast of Idaho Falls, Idaho, in Bonneville County.

Authorization: The Secretary of the Interior initially authorized the Palisades Project on December 9, 1941, under the provisions of Section 9 of the Reclamation Project Act of 1939 (53 Stat. 1187). Congress reauthorized the Palisades Project on September 30, 1950 (Public Law 864), 81st Congress, in accordance with a supplemental report approved by the Secretary in 1949.

Description: Palisades Powerplant serves large irrigation pumping power requirements on and near the Minidoka Project in southern Idaho.

History: Palisades Power Plant was completed in 1957, with the last unit going online in 1958. Due to Corona discharge the original windings lasted less than ten years. The generators were rewound in the 1960s, increasing the plant capacity to 130 mW. By the late 1980s the windings were again deteriorated. To take advantage of the large water releases for irrigation requirements the generators were rewound and uprated in 1991 through 1995 using class "H" winding insulation. The plant capacity was increased to its current value of 176 mW. This increased the hydraulic capacity of the powerplant to 10,000 cubic feet per second. Automation of the plant was started in 1995, with remote operation from Black Canyon Control Center beginning in April of 1999.

Parker Powerplant

Lower Colorado Region
Parker-Davis Project

Location: Parker Powerplant is located on the Colorado River about 155 miles downstream from Hoover Dam on the California-Arizona state line, twelve miles northeast of Parker, Arizona.

Authorization: The Parker-Davis Project was formed by consolidating the Parker Dam Power Project and the Davis Dam Project under the terms of the Act of May 28, 1954 (68 Stat. 143). The Parker Dam Power Project was authorized by the Rivers and Harbors Act of August 30, 1935 (49 Stat. 1028).

Description: The electrical integration and interconnection of Davis, Hoover, and Parker Powerplants provide maximum generation of power with efficient use of water resources. The highly developed agricultural base and the complex industrialization of the Pacific Southwest benefit greatly from Colorado River hydroelectric energy.

History: Parker Dam is a concrete arch structure with a volume of 380,000 cubic yards of concrete. The dam is 320 feet high and 856 feet long at its crest. Parker Dam impounds Lake Havasu, which includes a desilting basin and forebay from which the Metropolitan Water District of Southern California pumps water into its Colorado River Aqueduct. The Central Arizona Project also uses the forebay to pump Colorado River water into its canal system. With funds advanced by the Metropolitan Water District of Southern California, Reclamation awarded contracts for the construction of Parker Powerplant beginning in 1939. The Metropolitan Water District of Southern California, by virtue of advancing the funds for constructing Parker Dam and Powerplant, enjoys a perpetual contract with Reclamation, which provides that one-half of the powerplant's output is available to be used for the pumping of water into the aqueduct.

Pilot Butte Powerplant

Great Plains Region
Pick-Sloan Missouri Basin Program

Location: The Riverton Unit is in central Wyoming on the Wind River Indian Reservation. Pilot Butte Powerplant is located in Fremont County, Wyoming, approximately twenty-two miles northwest of Riverton, Wyoming, at the drop from the Wyoming Canal to Pilot Butte Reservoir.

Authorization: The project was authorized for construction by the Secretary of the Interior on June 19, 1918, under the terms of the Indian Appropriation Act for 1919, and approved by Congress on May 25, 1918. By the Act of June 5, 1920, the project was placed under Reclamation's jurisdiction. On September 25, 1970, Public Law 91-409 reauthorized the project as the Riverton Unit of the Pick-Sloan Missouri Basin Program.

Description: Pilot Butte Powerplant was built to supply power to the project.

History: The Pilot Butte Powerplant, located at the drop from the Wyoming Canal to Pilot Butte Reservoir, was taken out of service in June of 1973. A new penstock was installed, and the units were placed back in service in June of 1990.

Pole Hill Powerplant

Great Plains Region
Colorado-Big Thompson Project

Location: Pole Hill Powerplant is located in Larimer County, Colorado, near Little Hell Canyon, ten miles east of Estes Park, Colorado.

Authorization: The President approved the Secretary of the Interior's finding of feasibility on December 21, 1937.

Description: Pole Hill is a single unit powerplant. The plant is served by one penstock that carries the water down 840 feet from the canal above. When the plant is off-line, water is diverted through an outlet creating a spectacular waterfall at the north end of Little Hell Canyon. A small diversion dam at the mouth of the canyon collects the water and brings it back into the project.

History: Pole Hill began operating in 1954.

Roza Powerplant

Pacific Northwest Region
Yakima Project

Location: Roza Powerplant is located in Yakima County, Washington, on the Roza Main Canal, about three miles northeast of Yakima, Washington.

Authorization: The President approved the Yakima Project on November 6, 1935.

Description: The Roza Powerplant provides power at the cost of production to operate pumps for the Roza Irrigation District. The Roza Irrigation District pumps water to 27,700 acres of their irrigation lands. Any other power generated is marketed through BPA.

History: Plans for the Roza Division were initiated in 1932. The powerplant was completed in conjunction with that plan in 1958. Since then, it has been operated at the maximum capacity allowed by available flows, weather, and maintenance.

San Luis Pump-Generating Plant

Mid-Pacific Region
Central Valley Project

Location: San Luis Pump-Generating Plant is located in Merced County, California, on the San Luis Creek, twelve miles west of Los Banos, California.

Authorization: Funds for construction of the initial features of the Central Valley Project were provided by the Emergency Relief Appropriation Act of 1935 (49 Stat. 115). The Secretary of the Interior authorized the project, and the President approved it on December 2, 1935. The San Luis Unit, West San Joaquin Division, was authorized as a part of the Central Valley Project on June 3, 1960 (Public Law 86-488).

Description: The San Luis Pump-Generating Plant pumps Central Valley Project water for off-stream storage.

History: The powerplant is operated and maintained by the state of California under an operation and maintenance agreement with Reclamation. The name of the pump-generating plant has been changed by the state of California to Willam R. Gianelli.

Seminoe Powerplant

Great Plains Region
Kendrick Project

Location: Seminoe Powerplant is located in Carbon County, Wyoming, on the North Platte River, about seventy-two miles southwest of Casper, Wyoming.

Authorization: The President approved the Kendrick Project on August 30, 1935. Originally known as Casper-Alcova, the project was renamed Kendrick in 1937.

Description: Electric energy generated at the Seminoe Powerplant is marketed through the Pick-Sloan Missouri Basin Program's integrated system.

History: Seminoe Dam and Powerplant construction began in 1936 and was completed in 1939. The original installed capacity of the plant was 32,400 kW, but the plant was uprated in the mid-1970s to its current installed capacity of 45,000 kW.

Shasta Powerplant

Mid-Pacific Region
Central Valley Project

Location: Shasta Powerplant is located on the Sacramento River in Shasta County, California, nine miles northwest of Redding, California.

Authorization: Funds for construction of the initial features of the Central Valley Project were provided by the Emergency Relief Appropriation Act of 1935 (49 Stat. 115). The Secretary of the Interior authorized the project, and the President approved it on December 2, 1935. The Shasta and Trinity River Division was authorized by Public Law 386, 84th Congress, 1st Session, approved August 12, 1955.

Description: The Shasta Powerplant is a peaking plant. Its power is dedicated first to meeting the requirements of the project facilities. The remaining energy is marketed to various preference customers in northern California.

History: Unit Nos. 1 and 2 were uprated to 125 mW each in 1980. Unit Nos. 3, and 5 were uprated from 1968-1974. Unit Nos. 3, 4, and 5 generators were uprated to 142 mW in 2000, 1998, and 1999, respectively. Unit Nos. 3, 4, and 5 turbines presently restrict output to 125 mW for each unit.

Shoshone Powerplant

Great Plains Region
Shoshone Project

Location: Shoshone Powerplant is near the base of Buffalo Bill Dam. Shoshone Powerplant is located in Park County, Wyoming, approximately four and a half miles southwest of Cody, Wyoming, on the Shoshone River.

Authorization: The Secretary of the Interior authorized the Shoshone Project on February 10, 1904, under authority of the Reclamation Act of June 17, 1902.

Description: Power produced on the project is fed into a grid system, which serves an area extending into three states.

History: Shoshone Powerplant began operation in 1922, and the third unit came online in 1931. The original total installed capacity of the three units was 6,012 kW. In 1980, the plant was shut down due to the deteriorated condition of the units. Unit Nos. 1 and 2 remain in place in a decommissioned status. Unit No. 3 was removed and replaced with a new 3,000 kW unit in 1991.

Spirit Mountain Powerplant

Great Plains Region
Shoshone Project

Location: Spirit Mountain Powerplant is located in Park County, Wyoming, approximately four miles southwest of Cody, Wyoming, on a conduit from Buffalo Bill Dam.

Authorization: The Secretary of the Interior authorized the Shoshone Project on February 10, 1904, under authority of the Reclamation Act of June 17, 1902.

Description: The primary purpose of the Spirit Mountain Powerplant is to dissipate energy of the water from the pressurized supply conduit from Buffalo Bill Reservoir before it enters the unpressurized supply canal which feeds the Heart Mountain Powerplant and the Heart Mountain Canal. The secondary purpose of the powerplant is to produce electrical power.

History: Operation of this plant began in October of 1994.

Spring Creek Powerplant

Mid-Pacific Region
Central Valley Project

Location: Spring Creek Powerplant is located in Shasta County, California, on the Spring Creek arm of Keswick Reservoir near Redding, California.

Authorization: Funds for construction of the initial features of the Central Valley Project were provided by the Emergency Relief Appropriation Act of 1935 (49 Stat. 115). The Secretary of the Interior authorized the project, and the President approved it on December 2, 1935. The Shasta and Trinity River Division was authorized by Public Law 386, 84th Congress, 1st Session, approved August 12, 1955.

Description: Spring Creek Powerplant is at the foot of Spring Creek Debris Dam. The dam is an earthfill structure, 196 feet high with a crest length of 1,110 feet. Water for power is received through Spring Creek Tunnel, which diverts water from Whiskeytown Lake on Clear Creek. Water from the plant is discharged to Keswick Reservoir.

History: These facilities were built and are operated by Reclamation.

Stampede Powerplant

Mid-Pacific Region
Washoe Project

Location: Stampede Powerplant is located in Sierra County, California, on the Little Truckee River. Stampede Dam and Reservoir are located immediately below the mouth of Davis Creek and approximately eight miles above the confluence of the Little Truckee and Truckee Rivers.

Authorization: The project was authorized by Public Law 858, 84th Congress, 2nd Session, August 1, 1956, as amended August 21, 1958 (Public Law 85-706).

Description: Stampede Powerplant is a run-of-the-river plant. The power generated is dedicated first to meeting the requirements of the project facilities. The remaining energy is marketed to various preference customers in northern California.

History: These facilities were built and are operated by Reclamation. The powerplant was placed online in 1988.

Towaoc Powerplant

Upper Colorado Region
Dolores Project

Location: Towaoc Powerplant is located on the Towaoc Canal,five miles north of Cortez, Colorado, in Montezuma County, Colorado.

Authorization: The Dolores Project was authorized by the Colorado River Basin Act of September 30, 1968 (Public Law 90-537), as a participating project under the Colorado River Storage Project Act of April 11, 1956 (Public Law 84-485).

Description: Towaoc Powerplant provides energy to project facilities.

History: The powerplant is operated and maintained under a contract with the Dolores Water Conservancy District. The switchyard is maintained under a contract with Empire Electric Association, Inc.

Mid-Pacific Region
Central Valley Project

Location: Trinity Powerplant is located on the Trinity River in Trinity County, California, nine miles upstream from Lewiston, California.

Authorization: Funds for construction of the initial features of the Central Valley Project were provided by the Emergency Relief Appropriation Act of 1935 (49 Stat. 115). The Secretary of the Interior authorized the project, and the President approved it on December 2, 1935. The Shasta and Trinity River Division was authorized by Public Law 386, 84th Congress, 1st Session, approved August 12, 1955.

Description: Trinity Powerplant is a peaking plant. The power generated is dedicated first to meeting the requirements of the project facilities. The remaining energy is marketed to various preference customers in northern California.

History: These facilities were built and are operated by the Bureau of Reclamation.

Upper Molina Powerplant

Upper Colorado Region
Collbran Project

Location: The Upper Molina Powerplant is located in Mesa County, Colorado, on the east bank of Cottonwood Creek, about seven miles southeast of Molina, Colorado, and about twenty-three miles northeast of Palisade, Colorado.

Authorization: Congress authorized the Collbran Project on July 3, 1952, under Public Law 445, 82nd Congress, 2nd Session.

Description: Electrical energy is generated for use in west-central Colorado.

History: In 1992, the unit control system was upgraded. A UPS system, static exciter, SCADA, and Sy/Max 400 PLC were installed at Upper Molina Powerplant. This reconfigured the plant from local operation to remote.

Great Plains Region
Pick-Sloan Missouri Basin Program

Location: Yellowtail Dam, Powerplant, and Afterbay Dam, the principal structures of the Yellowtail Unit, are located on the Bighorn River, within the Crow Indian Reservation in Big Horn County in south-central Montana. The nearest town of consequential size is Hardin, Montana, approximately forty-five miles to the northeast.

Authorization: The Flood Control Act of December 22, 1944 (Public Law 534), authorized construction of Yellowtail Dam.

Description: Yellowtail Powerplant, with an installed capacity of 250,000 kW, was constructed as a multipurpose development providing low cost power generation, irrigation, and flood control. Facilities consist of Yellowtail Dam, Bighorn Lake on the Bighorn River, Yellowtail Powerplant at the toe of the dam, Yellowtail Afterbay Dam a short distance downstream, the government camp, and related facilities.

History: Construction began on Yellowtail Dam and Powerplant in May of 1961 and was completed in December of 1967. Construction of Yellowtail Afterbay Dam was started in April of 1964 and completed in November of 1966.